BIOMECHANICS OF THE HAND

BIOMECHANICS OF THE HAND

A Basic Research Study

Edmund Y S Chao, *Ph.D.*
Kai-Nan An, *Ph.D.*
William P Cooney III, *M.D*
Ronald L Linscheid, *M.D*

Biomechanics Laboratory
Department of Orthopedic Surgery
Mayo Clinic/Mayo Foundation
Rochester, Minnesota 55905
U S A

World Scientific
Singapore • New Jersey • London • Hong Kong

Published by

World Scientific Publishing Co. Pte. Ltd.
P O Box 128, Farrer Road, Singapore 9128

USA office: World Scientific Publishing Co., Inc.
687 Hartwell Street, Teaneck, NJ 07666, USA

UK office: World Scientific Publishing Co. Pte. Ltd.
73 Lynton Mead, Totteridge, London N20 8DH, England

Library of Congress Cataloging-in-Publication Data

    Biomechanics of the hand.
1. Hand. 2. Human mechanics. I. Chao, E. Y.
QP310.H36B56   1989   612'.97   87-28018
ISBN 9971-50-103-1
ISBN 9971-50-104-X (pbk.)

Printed in Singapore by JBW Printers & Binders Pte. Ltd.

PREFACE

Together with the brain, the hand is the most important organ for accomplishing tasks of adaptation, exploration, prehension, perception and manipulation, unique to humans. The importance of the hand to human culture is emphasized by its depiction in art and sculpture, its reference frequency in vocabulary and phraseology, and its importance in communication and expression. While we have accumulated an impressive understanding of the functional anatomy of the hand, our present state of knowledge seems well behind where current technology can lead us. This delay is not due to lack of recognition but rather the absence of scientific methodology for studying musculoskeletal entities in a quantitative and objective fashion. The development of a new discipline — Biomechanics — in its modern form, occurring in the last two decades provides us with the technical capability to accomplish new goals.

The functional uniqueness of the human hand was recognized and discussed by Aristotle and Hippocrates. Galen described well the human osteology and, anticipating our modern evolutionary interest, substituted a detailed dissection of the rhesus forelimb and hand for his contribution to myology. Empiric interest in the anatomy and function of the hand was revived during the Vesalian period. A particularly prescient work occurred in 1833 as the Bridgewater Treatise by Sir Charles Bell. Entitled **"The Hand — Its Mechanism and Vital Endowments as Evincing Design"**, Bell's foresight proclaimed, *"Some may conceive that as I have for my title, the Human Hand, and the relation of the solid structures of the animal frame, it will lead me to consider the body as a machine only. I neither see the necessity for this, nor do I acknowledge the danger of considering it in that light..... I cannot see how skepticism should arise out of the contemplation of the structure and mechanism of the animal body."* Intentional or unintentional, Sir Charles laid the cornerstone for modern biomechanics.

Recognizing that scientific inquiry involving biological systems has specific limitations, many scientists have similarly sought to combine anatomic observation with mechanical principles. Newer insights regarding anatomy and

v

pathology alone are less frequent, while anatomical and clinical pathomechanic studies have provided few new perspectives on hand function. Thus, when studying the unique endowments of the human hand, a fresh approach was sought. The present study was undertaken in an attempt to apply a composite approach to normal and pathological hand function by utilizing mechanical modeling, anatomical testing and clinical verification.

Our studies are not unique among those of the past or contemporary investigators, but they have the special flavor of combining collaborative efforts of the disciplines of orthopedic surgery and engineering science. Prior efforts, such as that by Duchenne, involved observation only. His monumental treatise on the electrophysiologic response of human muscle provides deep insight through descriptions of forearm and hand movements in relation to individual muscular contractions and helped to advance traditional descriptive anatomy to modern functional anatomy. Despite the progress of subsequent investigations, the complexity of the musculoskeletal aspects of the hand prevented a comprehensive analysis of function. While many innovative and clinically related descriptions of the hand joints and musculotendinous units have been established by Bunnell, Littler, Kaplan, Landsmeer, Flatt and Fischer, Brand, and others, the inability to quantitatively describe the complex geometry and to solve the internal muscle and joint forces as an indeterminate problem have persisted. The need for sophisticated biomechanical studies has been propelled by the demand for more objective evaluation of the ever-increasing number of clinical procedures which have evolved. Diagnostic, surgical and rehabilitative methods in the treatment of hand problems are often predicated on refined intuitive leaps across gaps formed between documented information and scientific rationales. Biomechanical studies on the function of the hand can help to close these gaps.

Precise observations, down to the microscopic level, have provided an increasing appreciation of the anatomic details of the structure of the hand. However, in modeling the entire structure of the hand, macroscopic analysis is necessary, since the musculoskeletal joint system must be envisioned as an entity. Although experiments in two-dimensional modeling have added considerably to the understanding of the more basic functional behavior of the structure, inherent anatomic and physiological complexities have defied such simplified methods. Newer techniques, such as computer-aided modeling capable of providing three-dimensional perspectives, will be more advantageous for generating realistic information on both normal and pathological hand mechanics.

Progress in orthopedics has been substantial in recent decades, and these advances have not excluded the hand. New treatment modalities are frequently evolved from research and new research has been stimulated by treatment failures. While the concept of total joint replacement has set forth a dramatic backdrop for this progress, it has also highlighted the need for objective information on joint mechanics. Such needs have stimulated the rapidly expanding musculoskeletal specialties in which biomechanics has found a natural habitat.

Collaboration between medical specialties and engineering sciences appears to provide the most effective means of generating refined information on joint function. A potentially fruitful method within this framework is the idealization of biologic systems as equivalent mechanical models, which may be subjected to well-established engineering analysis. Many clinical conditions do not lend themselves to direct specimen experimentation, but proper simulation based on analytical models can fulfill the investigatory goals in an efficient manner. Pertinent physiologic principles and anatomic constraints can be incorporated in such models. The orthopedic surgeon can play a key role in safeguarding the validity of such analyses, but a basic science approach is needed to accomplish them. We should recognize that this approach is but one of many which may be employed in musculoskeletal analysis. Validity is assured by a careful balance between anatomical, clinical and biomechanical investigations.

The progressive development of the hand and its unique relationship with the central nervous system appears to have been intrinsically related to hominid evolution. The gift of opposability of the thumb to the fingers endowed man with precision pinch, a significant refinement to the crude prehension of our arboreal forebears. The strength and precision of these manipulative abilities remain the basic determinants for communicating with our environment and one another. The study of the anatomic structures and their functions is important, not only for application to the diseased and injured hand, but also in design of tools for modern society.

It is said that the practice of surgery is more art than science. If this were true, only the few with a gift of skill and ingenuity could reach a sufficient level of competence. The knowledge and expertise necessary to learn sophisticated techniques and exercise good judgment require a solid background of understanding in the basic sciences. Biomechanics can assist in accomplishing this by synthesizing anatomic, physiologic and pathologic problems in a rational format.

The present work obviously is not comprehensive and we recognize the mechanistic view does not suffice for every aspect of the structure and function of the hand. Since its inception ten years ago, this investigation has been balanced between orthopedic surgeons and biomechanicians to achieve reference. We have limited our effort to understanding the function of the hand, as related to the force and motion potentials observed in normal and pathological states. The methodologies adopted and data obtained thus far mark only the beginning of a long pursuit of a complicated mechanism. If our contribution is minute, as measured on the scale of progress, we hope the systematic approach to investigation of an anatomical system will aid our successors.

When Newton first published his laws of gravity, many contemporary philosophers cast doubt on what seemed an irrational and paradoxical thesis. They criticized him for his presumptions which contradicted the spiritual view of the universe then held by the intellects. In retort, Newton acknowledged, *"There is a wide difference between the ignorance of the child or of the peasant, and the consciousness of the philosopher that he has arrived at a point beyond which man's facilities do not carry him."* The same argument can be extended to the view on the mythical power of the hand. Sir Charles Bell has stated well, *"To have the mind awaken to the many proofs of design in the hand — to be brought to the conviction that every thing is orderly and systematic in its structure — that the most perfect mechanism, the most minute and curious apparatus, and sensibilities the most delicate and appropriate, are all combined in operation that we may move the hand."* It is in this spirit that we proceed forward to investigate the function of the hand. Although Newton may not have fully understood how his arm moved, he would have approved of utilizing the laws of mechanics to explain the power of the hand.

ACKNOWLEDGEMENTS

This study has been supported in its entirety by NIH Grant no. AM17172. Public acknowledgment is gratefully made to the following who have contributed significantly to the work reported here.

Jasper R. Daube, M.D.
R. Michael Gross, M.D.
Michael J. Lucca, M.D.
Kazuhisa Takahashi, M.D.
Linda J. Askew, R.P.T.
Fred E. Axmear, M.S.E.E.
Timothy P. Harrigan, Ph.D.

J. Donald Opgrande, M.D.
Jan P. DeRoos, M.D.
Yasuo Ueba, M.D.
Akio Minami, M.D.
Ross R. Hoffman, M.S.M.E.
Roberta A. Kasman, M.S., M.D.
B. Patrick Cahill, B.S.E.E.

The manuscript was carefully typed and meticulously put together in superb quality by Sharon Wellik and Joanne Murray. We wish to express our special gratitude to them. Other members of the Biomechanics Laboratory and the staff of the Department of Orthopedic Surgery have kindly assisted and encouraged us in this study. This report truly represents a collaborative effort between the medical and engineering professions.

CONTENTS

INTRODUCTION

This book contains selected works on biomechanics of the hand accumulated in the Biomechanics Laboratory of the Mayo Clinic during the past eleven years. Though we intended to document the methods, materials and data in a comprehensive manner, some degree of abbreviation was necessary to facilitate easier reading by the general audience. Certain detailed information purposefully omitted from the present text may be supplemented by existing publications of our work. These publications are cited at the end of each chapter for convenience of reference if more extensive deliberation is desired. Each chapter included in this manuscript can stand alone without referring to the remaining chapters.

Chapter 1 presents the normative models of each digit in the hand for the purpose of performing muscle and joint force analyses. These models were derived from carefully performed anatomic and radiographic experiments based on multiple specimens. Tendon orientation and its contribution to the analytical model in terms of force and moment parameters are incorporated in the governing equilibrium equations. The moment arm data at each joint was determined by the radiographic technique and secondarily validated by an adjuvant method using tendon excursion. Such data were also used to demonstrate the changes of moment arms as a function of joint angulation. Although most tendons and intrinsic muscles of the hand are axially oriented to provide strong flexion and extension functions, they often also have a component of force producing rotation or deviation. In particular, the flexors often tend to deviate the proximal phalanx toward the ulnar side at the metacarpophalangeal (MP) joint of the fingers. This produces a significant ulnar deviation if the joint is unbalanced or unstable. The extensors seemed to have a constant moment arm throughout the range of joint flexion, but the flexors exhibited a changing moment arm pattern at different joint positions. The extrinsic tendon and intrinsic muscle moment arms are now defined throughout their ranges of motion, and force analysis can be performed at different functional configurations of the hand.

Chapter 2 summarizes the cross-sectional study of both the intrinsic and extrinsic muscles of the hand. The morphometric data include the muscle cross-sectional area and centroids at different levels of the forearm and hand, muscle centroidal line (resultant force vector) relative to the bone and the joint, and the volume of each muscle and its length from origin to insertion. A total of eleven normal specimens were carefully studied, and in addition, seven specimens were dissected so that each muscle's mean fiber length could be measured and averaged. The physiological cross-sectional area (PCSA) for each hand muscle was calculated and defined as the volume of that muscle divided by its mean fiber length. Muscle PCSA's were then combined with moment arm information to assess functional potentials of rotating joints in three mutually perpendicular planes (flexion-extension, abduction-adduction or radial-ulnar deviation, and supination-pronation). The PCSA information was used to constrain muscle force prediction in the theoretical analysis, and the functional potential data were applied to speculate on the contributions of different muscles in rotating each joint when they are maximally contracted. These data are useful in studying joint balance and in the consideration of tendon transfers due to peripheral nerve paralysis affecting either the intrinsic or extrinsic muscles.

Chapter 3 provides the basic methods used to determine internal forces in the hand during isometric functions. Since the number of unknown variables in the model exceeds that of the equilibrium and constraint equations, the well-known indeterminate problem resulted. The problem was solved by the systematic reduction method and the optimization technique. Although the optimal criteria adopted in the optimization procedure were not yet justified experimentally, the current results provide a solution domain which circumscribes the potential level of function of the anatomic components involved in the system. It was found that the finger is capable of producing the same functional strength through numerous combinations of muscle activities if such strength is of a nominal magnitude. When a load becomes maximal, a unique combination of individual muscle forces may exist to accomplish the demanded task. Based on the normative models developed, the maximum tendon force in isometric pinch and grasp activities can reach five to twelve times the applied force. Joint contact forces are also quite high, particularly at the MP joint. On a relative basis, hand joints should be regarded as load-bearing joints, similar to those in the lower extremities. The same method has also been applied to obtain the internal force of the hand involved in other daily activities.

Chapters 4 and 5 summarize the method and results of hand joint orientation, range of motion, and functional strength in the normal population and in patients with hand problems before and after therapeutic and surgical treatments. A hand-held goniometer and a custom-made dynamometer were used to measure joint ranges of motion and selected strengths in the fingers and thumb. However, certain joints cannot be accurately and conveniently measured by the goniometer, particularly those under functional postures or with severe deformities. For this reason, a biplanar radiographic method was developed and used for the analysis of the carpometacarpal (CMC) joint at the base of the thumb. A surface hardness gauge was modified to measure skin compliance as a means to quantitate joint swelling and contracture. Prehensile manipulation of the hand was tested using a newly developed timed pegboard. Although over one hundred normal subjects have been studied, the data varies significantly, even after proper normalization based on anatomic measurements. It was concluded that categorizing the results according to sex, age and occupation is more appropriate to standardize a normative data base. Several groups of patients with severe hand abnormalities were studied to provide objective assessments of the effects of treatment. In these studies, the biomechanical evaluation was found to correlate more closely with the clinical results than with other laboratory tests. Distinguishable patterns of strength distribution among the fingers were observed for different types of peripheral neuropathies. A procedure is under development for different diagnosis of the peripheral nerve lesion.

Hand pathomechanics, as related to several common clinical entities, were simulated based on our analytical model, and the results are presented in Chapter 6. These abnormal conditions included muscle weakness caused by either peripheral nerve palsies or traumatic injuries, phalangeal fracture nonunion, and subluxation of tendons due to pulley breakdown. In addition, several surgical techniques, including tendon transfer and distal interphalangeal (DIP) joint fusion, were also analyzed. The predicted results appear to mimic the clinical observation which reflects the usefulness of the theoretical model in analyzing abnormal hand conditions. These results, after certain refinements, are expected to be useful in establishing treatment rationales.

Chapter 7 describes a semi-quantitative method of validating the theoretically predicted hand muscle forces based on direct electromyographical (EMG) assessment. Raw EMG signals of certain muscles under isometric hand strength tests were rectified and integrated. These processed EMG signals were correlated, using normative hand models, with the calculated muscle forces based on a nonlinear relationship. The thumb muscles in seven normal subjects were studied under a series of controlled isometric functions, and all the muscles involved were monitored simultaneously. Upon increasing functional strength, the EMG signal of certain muscles would also increase proportionally, which provides a qualitative validation between the analytical results and those measured and interpreted from the EMG data. Under submaximal isometric strength testing of pinch and grasp functions, the calculated muscle forces and their integrated EMG signals seemed to correlate linearly. The temporal distributions of muscle activities were found to be dependent upon the type of hand function and the level of functional strength.

Chapter 8 describes the capsuloligamentous structures around the metacarpophalangeal joint and their role in joint constraint and stability. The morphometric data were collected by gross dissection and observation from both outside and inside the cadaveric finger joint specimens. The locations of the origin, insertion and variation of the length throughout the joint range of motion of each ligamentous element were obtained by using biplanar radiography. The capsuloligamentous contributions to joint constraint were also obtained by using load-displacement tests on an MTS machine. The relative contribution of each ligament was obtained by repeating the test with sequential sectioning of the soft tissue elements. Four modes of displacement (distraction, volar-dorsal displacement, abduction-adduction and axial rotation) of the proximal phalanx were examined at both the extended and flexed positions. These data are useful in understanding the laxity and stability of the MCP joints.

The current work, pursued for the past eleven years, needs continued study and refinement in certain areas. The established research-oriented results need broadening for potential clinical applications. Hand functional evaluation and the simulation of pathological conditions are two such promising areas. The mechanical strength of tendons at various degrees of laceration, before and after surgical repair, is also being studied. An *in vivo* muscle force measurement technique, using a buckle transducer attached directly to tendons of patients undergoing carpal tunnel procedures, is being developed to provide direct verification of calculated muscle forces in isometric functions. These ongoing studies are yet unable to provide definitive conclusions and have purposefully been eliminated from this report. They will serve to guide the directions of our future research.

1

NORMATIVE MODEL OF HUMAN HAND

A three-dimensional normative model of the hand was established based on the anatomical study of normal hand specimens. In the first study, the joint and tendon orientations were defined from biplanar X-ray methods. Force potential and moment potential parameters were utilized to describe the contribution of each tendon in the force analysis. The mean values of these two parameters were used to compute the orientation of the tendons and intrinsic muscles at each joint in the normative model. With appropriate coordinate transformations at the joints, the tendon locations and excursions under various functional configurations can be derived. This model can be used to perform force and motion analyses for both normal and pathological hands. In the second study, tendon excursions during rotation of individual index finger joints were recorded continuously throughout the joints' ranges of motion. Both intrinsic and extrinsic muscles were studied during flexion-extension and abduction-adduction functions. Moment arms of tendons with respect to joint centers were further derived from excursion data to confirm the results of the normative model obtained using biplanar X-ray technique.

INTRODUCTION

The complexities of the function and anatomy of the human hand have long been recognized[2,6,8,10,11,15,16]. From a biomechanical standpoint, the human hand can be considered as a linkage system of intercalated bony segments. The joints between each phalanx are spanned by ligaments, tendons and muscles. With the contraction of muscles, these joints can be moved in a characteristic manner constrained by the interposing soft tissues and the bony articulation. If the motion is resisted, functional strength in the forms of pinch and grasp can result. In the hand, most of the tendons span the joint and continue their course over one or more joints, thus forming a bi-articular or poly-articular system.

The functional anatomy of the spatial relationships between these tendons and muscles and their associated joints has been extensively studied by Landsmeer[11,12]. He also proposed a series of models to represent the various manners in which tendons bridge the associated joints. Several biomechanical models of various aspects of hand function have been developed[7,14]. However, these studies show either a lack of quantitative description or information restricted to only two dimensions. The objective of the present study is to establish a workable model in a three-dimensional manner, based on the careful measurement of normal specimens.

The present model and the associated data can help other investigators who wish to pursue biomechanical analysis of the hand but have limited clinical facilities and resources for performing detailed anatomical studies. Such a model can easily be utilized in the study of hand motion or force analysis of the hand under various functional activities. Systematic variation of this model based on clinical examinations can provide an effective means to study the internal forces in the hand suffering pathological changes.

EXPERIMENTAL METHOD

A.   *Biplanar X-ray Technique*

In the study of the three-dimensional locations of tendons with respect to the bony segments at the joint, 15 fresh cadaver specimens were used. The experimental techniques are the same as those described previously[1,3]. By the use of standard surgical incisions, a small segment of each tendon and muscle near the joint was carefully exposed. Markers made of different grades of surgical wire with varying lengths were inserted into the center of each tendon and muscle at sites immediately proximal and distal to the joint. Finger and thumb specimens with wires embedded were then subjected to biplanar X-ray exposure on dental film. A specially designed holder for alignment of the X-ray beam and film with the dorsal-volar and radial-ulnar plane of the finger, or thumb, specimen was used so that accurate AP and lateral X-rays could be obtained. The proper position for the insertion of the markers was determined with reference to the pulley systems, retinacular ligaments and transverse bands, which constrain the tendons and muscles at the joint[4,5,10,13,17]. These constraint ligaments and pulleys minimize bowstringing with flexion and extension of the joints, such that a straight-line relationship exists along the tendon markers and best represents the direction of the tendon force. The use of different sizes and lengths of wires facilitated easy identification of the markers on the X-ray films. The selection of wire size depended on the diameter of the tendon, in order to make insertion possible. For flexor tendons, small incisions in the tendon sheath were made for the insertion of wires so that the pulley systems were not damaged. Careful closure was then performed. This procedure was repeated for all the tendons at their associated joints, as defined in

Table 1-1. Enlarged prints of the two X-ray films (Fig. 1-1) were then utilized for the analysis of the tendon locations for each joint.

Table 1-1   Tendons and muscles involved in hand function.

Hand Element	Joint	Unknown Tendon and Intrinsic Muscle Forces
Finger	DIP	Terminal extensor (TE) Flexor profundus (FP)
	PIP	Extensor slip (ES) Radial band (RB) Ulnar band (UB) Flexor sublimis (FS)
	MP	Long extensor (LE) Radial interosseous (RI) Ulnar interosseous (UI) Lumbrical (LU)
Thumb	IP	Flexor pollicis longus (FPL) Extensor pollicis longus (EPL)
	MCP	Abductor pollicis brevis (APB) Flexor pollicis brevis (FPB) Adductor pollicis (ADD) Extensor pollicis brevis (EPB)
	CMC	Opponens pollicis (OPP) Abductor pollicis longus (APL)

B. Tendon Excursion Study

Seven hand specimens from amputated limbs and cadavers were used in the experiment. The index finger was disarticulated at the CMC joint. Contact AP and lateral X-rays were taken to confirm the normality of the specimen and to obtain the anthropometric measurements as well. Among the seven specimens, three were right hands and four were left hands.

All of the extrinsic muscle tendons, including both the extensor digitorum communis (EC) and extensor indicis (EI), were exposed and incised at the level of the midshaft of the metacarpal. For the intrinsic muscles, the musculotendinous junctions were incised after removal of the skin and subcutaneous soft tissues around the metacarpals. Great caution was taken to avoid impairing the pulley mechanisms of all the tendons. All of the skin and soft tissue distal to the metacarpal neck was left intact. A 4-0 nylon suture was tied to the proximal end of each of the tendons.

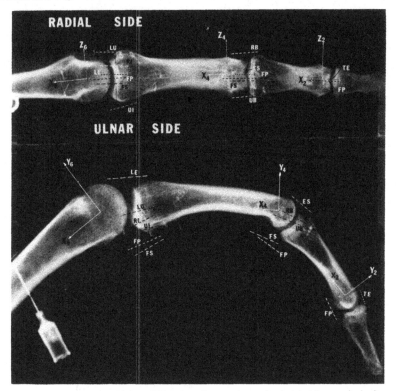

Fig. 1-1   Biplanar X-ray views of finger tendons and muscles located by surgical wires.

During testing, the bony segment proximal to the joint was clamped to the upright portion of the test frame. An electrogoniometer, made of a rotary potentiometer, was used to monitor joint angular displacement. Attachment of the goniometer to the bony segments adjacent to the joint was achieved through a gliding linkage, so that the connecting rods remained parallel to the bony shafts through the range of motion. For measurement of tendon excursion, another rotary potentiometer with a rotating drum was utilized. The nylon suture attached to the tendon was extended with a thin metal wire in the direction parallel to the metacarpal shaft. The metal wire was then wrapped around the drum of the rotary potentiometer. The other end of the wire was connected to a 500 g load. The linear displacement was then obtained by multiplying the radius of the drum and the angle of drum rotation. The goniometer and displacement transducer were calibrated to an accuracy of $\pm 1°$ and ± 0.1 mm, respectively. Electrical potentials from the displacement transducers and goniometers were simultaneously and continuously recorded on a Hewlett-Packard 7044 X-Y recorder.

The experiment was carried out by sequentially monitoring each tendon at each joint individually. Movement of other joints was prevented by splint stabilization in the neutral extended position. Angular motion of the joint was confined to one single plane. The finger was moved passively and slowly while the joint was allowed to rotate without constraint. The tendon was kept under constant tension, while equilibrium was maintained by passive resistance at the finger tip. Each test was repeated five times through the entire range of motion to ensure reproducibility of the experimental data. This excursion experiment was carried out for four of the finger point motions; namely, flexion-extension of the distal interphalangeal (DIP) joint, proximal interphalangeal (PIP) joint, and metacarpophalangeal (MP) joint in the abducted position, as well as abduction-adduction of the MP joint in the extended position.

DATA REDUCTION

A. *Formulation of the Model*

In order to establish a mathematically workable model, six Cartesian coordinate systems were established (Fig. 1-2). These coordinates were used to define the locations and orientations of tendons and can also be utilized to describe the joint configuration. There are two coordinate systems for both the middle and proximal phalanges and only one system for both the distal phalanx and metacarpal. The primary systems (numbers 2, 4 and 6) are located at the approximate center of rotation of the phalangeal and metacarpal heads, and the secondary systems (numbers 1, 3 and 5) are a translation of the proximal systems to the centers of the concave articular surfaces. The x-axis is projected along the phalangeal or the metacarpal shaft, passing from the center of rotation to the center of the concave articular surface at the proximal end. The y-axis is projected dorsally, and the z-axis is projected radially for the right hand and ulnarly for the left hand.

With the assistance of these coordinate systems, the tendon locations in three-dimensional space can be determined and measured. At each joint, the tendon is represented by a straight line joining two points designated by the markers; one on the distal side of the joint and the other on the proximal side of the joint. The three-dimensional locations of these two points on each tendon were expressed, based on the distal and proximal coordinate systems.

The coordinates for each marker in the 15 cadaver specimens were not used in the averaging process, since not all of the markers in the tendons for each specimen have identical positions, and this would introduce significant variations in the final data. Instead, two parameters used to describe the

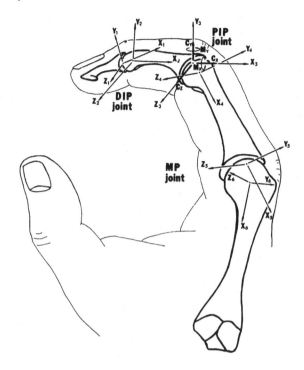

Fig. 1-2   Coordinate systems used to define tendon locations.

orientation and location of each tendon for the final model were determined first from raw data. These two parameters are the 'force potential' and 'moment potential'. The force potential is expressed in terms of the directional cosine of a tendon with respect to the distal system, and the treatment potential specifies the moment arm of the tendon with regard to the joint center and in the direction of each coordinate axis of the distal system. Obviously, force potential provides the contribution of a particular tendon in generating joint constraint forces, and the moment potential specifies the functional moment of each tendon in rotating the joint at three mutually perpendicular directions. The mean values of these parameters among the 15 specimens were then used to construct the normative model.

The application of force and moment potentials can be demonstrated by the static force analysis of the hand under isometric functions. In deriving the equilibrium equations about the center of rotation of the joint, the following forms are usually used:

Force equations

$$\Sigma\alpha_i F_i + C_x + R_x = 0,$$
$$\Sigma\beta_i F_i + C_y + R_y = 0, \qquad (1)$$
$$\Sigma\gamma_i F_i + C_z + R_z = 0.$$

Moment equations

$$\Sigma a_i F_i + M_x + T_x = 0,$$
$$\Sigma b_i F_i + M_y + T_y = 0, \qquad (2)$$
$$\Sigma c_i F_i + M_z + T_z = 0,$$

where

$$\begin{aligned}
\alpha_i, \beta_i, \gamma_i \quad &= \text{force potential parameters,} \\
a_i, b_i, c_i \quad &= \text{moment potential parameters,} \\
C_x, C_y, C_z \quad &= \text{unknown joint constraint forces,} \\
M_x, M_y, M_z \quad &= \text{unknown joint constraint moments,} \\
F_i \quad &= \text{unknown tendon or muscle forces,} \\
R_x, R_y, R_z \quad &= \text{externally applied forces, and} \\
T_x, T_y, T_z \quad &= \text{externally applied moments.}
\end{aligned}$$

Based on the parameters in the model, static force analysis can easily be applied for the determination of the unknown tendon forces and joint constraint forces and moments.

The coordinates of two points on each tendon at the joint were determined from the averaged values of force and moment potential parameters (Fig. 1-3). Several steps were involved in this reduction procedure. First, the force potential parameters were normalized such that these parameters of each tendon form a unit force vector. Second, the average moment potential parameters were adjusted, if necessary, so that the moment arm and unit force vector of each tendon were orthogonal. Finally, these normalized and adjusted force and moment potential parameters were ready for calculating the coordinates of points for tendons. Each pair of points was expressed with respect to the distal and proximal coordinate system at each joint. It was assumed that the tendons were constrained to remain fixed in relationship with the proximal and distal coordinate systems by pulleys and ligaments. Therefore, the results obtained at the neutral position could later be used for any finger configuration with the appropriate coordinate transformations (Fig. 1-3).

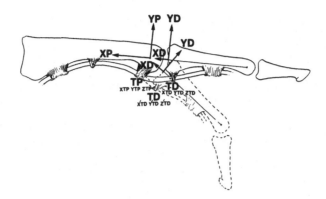

Fig. 1-3 Two points on each tendon of the joint have fixed relationships with their coordinate systems.

When the finger is in a functional configuration other than the neutral position, the proximal coordinate system at a particular joint can be related to the distal system through simple rotational and translational transformation. The orientation between two digits was defined by classic Eulerian angles. These angles are flexion-extension (ϕ), radio-ulnar deviation (θ), and axial rotation (ψ). The rotational sequence of these angles, as related to both the distal (moving system) and proximal (fixed system) coordinate systems at a joint is illustrated (Fig. 1-4). Using these Eulerian angles, the coordinates of a tendon point or the components of a force vector, as defined in either system, can be related by the following transformation equation:

$$
\begin{vmatrix} X_D \\ Y_D \\ Z_D \end{vmatrix} = \begin{vmatrix} c\theta c\phi & c\theta s\phi & -s\theta \\ -c\psi s\phi + s\psi s\phi c\theta & c\psi c\phi + s\psi s\theta s\phi & s\psi c\theta \\ s\phi s\psi + c\psi c\phi s\theta & -s\psi c\phi + c\psi s\theta s\phi & c\psi c\theta \end{vmatrix} \begin{vmatrix} X_p \\ Y_p \\ Z_p \end{vmatrix} + \begin{vmatrix} X_o \\ Y_o \\ Z_o \end{vmatrix} \qquad (3)
$$

in which

X_D, Y_D, Z_D = coordinates of a tendon point or components of a vector measured with respect to the distal system,

X_p, Y_p, Z_p = coordinates of a tendon point or components of a vector measured with respect to the proximal system,

X_o, Y_o, Z_o = coordinates of the origin of the proximal system expressed in the distal system,

s = sine, and

c = cosine.

This transformation equation was utilized in the force analysis to express the tendon and externally applied force vectors in the appropriate coordinate system where the free body was taken.

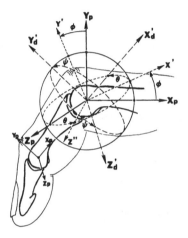

ϕ : Flexion-Extension (1st rotation — Z_p)
θ : Radioulnar Deviation (2nd rotation — Y')
ψ : Pronation-Supination (3rd rotation — X_d')

Fig. 1-4 Eulerian angles used to define the orientations of finger digits.

B. Tendon Excursion and Moment Arm

The procedures involved in reducing the tendon excursion data were straightforward. With appropriate calibration of the transducers, the value of excursion was derived from the recorded joint rotation-tendon displacement graphs. Excursion variations at various joint positions were visually displayed. However, if additional information on moment arms was desired, further reduction of excursion and joint displacement data was necessary.

The relationship between tendon excursions, joint displacements and moment arms has been demonstrated by Brand, *et al.*[16]. The exact moment arm of a tendon with respect to the axis of joint movement can be derived if the tendon can be exposed proximally to measure its excursion as the finger moves about one axis in a single joint. More specifically, the instantaneous moment arm of the tendon in the plane of motion at a specific joint configuration can be obtained from the slope of the graph of the tendon excursion versus joint rotational displacement. This relationship exists regardless of whether the graph is linear or nonlinear[19]. In addition, based on the three models of Landsmeer[12] and the normative model previously reported[1], the formulas for calculating instantaneous moment arms (M) by the slope of the excursion (E) versus joint angle (ϕ) were derived.

Model I

This model describes the situation in which tendons follow the curvature of the articular surface.

$$E = r \cdot \phi$$
$$M = r$$
$$dE/d\phi = r = M$$

where r is the radius of curvature of the articulating surface and d is the traditional symbol for differentiation.

Model III

In this model, the tendon runs in a tendon sheath which holds it firmly in a constant position against the shaft of the bone but allows the tendon to curve smoothly, conforming to the joint architecture.

$$E = 2y + \phi \cdot d' - \phi \cdot y/\tan(\phi/2)$$

$$M = h = d' - y/\tan\left(\frac{\phi}{2}\right) + y/\sin(\phi/2)$$

$$dE/d\phi = d' - \frac{y}{\tan\left(\frac{\phi}{2}\right)} + \frac{y}{\sin\left(\frac{\phi}{2}\right)} \cdot \frac{\phi/2}{\sin\left(\frac{\phi}{2}\right)}$$

$$dE/d\phi = M, \text{ when } \sin\left(\frac{\phi}{2}\right) \simeq \frac{\phi}{2}$$

The definitions for y, h and d' are illustrated in Fig. 1-5.

In order to perform the above calculation, digitization of the recorded curves of tendon excursions versus joint angles was required. A Graf/Pen sonic digitizer was utilized for this purpose, and the data were stored in a digital computer. Quintic spline function[2] was used for smoothing and curve

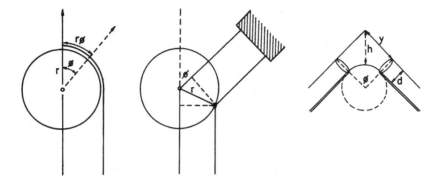

Fig. 1-5 Landsmeer's two-dimensional models of tendon bridging a joint.

fitting the raw data; then the first derivative of the tendon excursion with respect to the joint angle was obtained. The values of the derivative, which were equivalent to the moment arms at a particular joint angle, were tabulated and graphically displayed.

RESULTS

The linkage structure of the bony elements was related to the coordinate systems. The distance between the origins of two coordinate systems, i.e. the

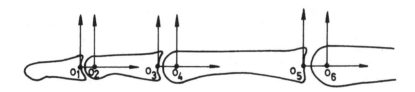

Distances Between Coordinate Systems
(mean and standard deviations)

	O_1O_2	O_2O_3	O_3O_4	O_4O_5	O_5O_6
Thumb	0.243 ± 0.035	1.0 ± 0	0.338 ± 0.042	1.424 ± 0.097	0.333 ± 0.047
Index	0.224 ± 0.034	1.0 ± 0	0.288 ± 0.036	1.919 ± 0.227	0.432 ± 0.058
Middle	0.184 ± 0.032	1.0 ± 0	0.233 ± 0.041	1.608 ± 0.087	0.365 ± 0.052
Ring	0.166 ± 0.031	1.0 ± 0	0.224 ± 0.041	1.576 ± 0.068	0.346 ± 0.030
Little	0.230 ± 0.042	1.0 ± 0	0.304 ± 0.046	1.797 ± 0.145	0.506 ± 0.082

Fig. 1-6 Distances between coordinate systems. Mean and standard deviations.

center of rotation and the center of the concave articular surface of the joint, was used to specify the location of the six coordinate systems. Figure 1-6 shows the average and the standard deviation of these distances. The dimensions are all normalised with respect to the distance between O_2 and O_3; that is, the distance between the centers of rotation of the DIP and the PIP joints.

In the analysis of data among 15 specimens, the exact tendon locations with respect to each localized coordinate system were expressed in dimensionless terms by normalizing their coordinates according to the length of the middle phalanx of the finger, to minimize anthropometric variations. In the case of a thumb, the length of the proximal phalanx was used as a common denominator.

The two representative points of each tendon at its associated joint were summarized in the dimensionless form in Tables 1-2a to 1-2e. As mentioned previously, the distal and proximal points of a tendon are measured with respect to the distal and proximal coordinate systems at the joint, respectively.

Table 1-2a   Tendon locations in thumb (mean value of five specimens).

Joint	Tendon	Distal Point			Proximal Point		
		X	Y	Z	X	Y	Z
IP	EPL	0.000	0.192	−0.057	−0.050	0.201	−0.044
	FPL	−0.007	−0.224	0.049	0.100	−0.318	0.034
MP	EPL	0.000	0.280	−0.157	−0.050	0.247	−0.224
	FPL	−0.062	−0.318	0.009	0.100	−0.521	−0.012
	ADD	−0.062	−0.224	−0.280	0.100	−0.575	−0.346
	EPB	0.000	0.265	0.057	−0.050	0.268	−0.019
	FPB	−0.062	−0.094	0.285	0.100	−0.416	0.435
	APB	−0.062	0.007	0.288	0.100	−0.160	0.533
CMC	EPL	0.000	0.179	−0.385	−0.050	0.225	−0.076
	FPL	−0.067	−0.476	−0.046	0.100	−0.152	−0.276
	ADD	−0.067	−0.065	−0.395	0.100	0.269	−0.589
	EPB	0.000	0.302	0.029	−0.050	0.184	0.132
	FPB	−0.067	−0.351	−0.284	0.100	0.004	−0.848
	APB	−0.067	−0.140	1.098	0.100	−0.273	0.410
	OPP	−0.067	−0.236	0.090	0.100	−0.493	−0.074
	APB	−0.067	0.346	0.212	0.100	0.136	0.239

Table 1-2b Tendon locations in index finger (mean value of 15 specimens).

		Distal Point			Proximal Point		
Joint	Tendon	X	Y	Z	X	Y	Z
IP	TE	0.004	0.199	−0.010	0.000	0.196	−0.009
	FDP	0.004	−0.184	0.026	0.300	−0.245	0.054
PIP	FDP	−0.212	−0.308	0.009	0.400	−0.409	0.027
	RB	−0.112	0.186	0.223	0.100	0.181	0.268
	UB	−0.112	0.151	−0.290	0.100	0.131	−0.312
	FDS	−0.212	−0.249	0.015	0.400	−0.311	0.028
	ES	−0.038	0.278	−0.027	0.000	0.266	−0.026
MP	FDP	−0.118	−0.386	0.031	0.300	−0.619	0.004
	FDS	−0.118	−0.477	−0.074	0.300	−0.689	−0.114
	RI	−0.318	−0.033	0.443	0.400	−0.362	0.629
	LU	−0.318	−0.148	0.370	0.400	−0.704	0.541
	UI	−0.318	−0.039	−0.461	0.400	−0.379	−0.442
	LE	−0.018	0.421	−0.033	0.000	0.483	−0.026

Table 1-2c Tendon locations in middle finger (mean value of 15 specimens).

		Distal Point			Proximal Point		
Joint	Tendon	X	Y	Z	X	Y	Z
IP	TE	−0.036	0.157	−0.014	0.000	0.169	−0.017
	FDP	−0.036	−0.158	0.054	0.300	−0.239	0.050
PIP	FDP	−0.267	−0.308	0.009	0.400	−0.251	−0.007
	RB	−0.167	0.206	0.237	0.100	0.132	0.262
	UB	−0.167	0.132	−0.262	0.100	0.079	−0.290
	FDS	−0.267	−0.217	0.054	0.400	−0.248	0.023
	ES	−0.017	0.241	−0.019	0.000	0.234	−0.009
MP	FDP	−0.317	−0.334	0.009	0.300	−0.522	0.001
	FDS	−0.185	−0.355	0.065	0.300	−0.593	0.063
	RI	−0.185	0.011	0.340	0.400	−0.499	0.491
	LU	−0.385	−0.174	0.311	0.400	−0.680	0.403
	UI	−0.385	0.011	−0.135	0.400	−0.185	−0.119
	LE	−0.085	0.352	−0.015	0.000	0.416	−0.013

Table 1-2d   Tendon locations in ring finger (mean value of 15 specimens).

Joint	Tendon	Distal Point			Proximal Point		
		X	Y	Z	X	Y	Z
IP	TE	−0.054	0.141	−0.016	0.000	0.154	−0.021
	FDP	−0.054	−0.160	0.001	0.300	−0.239	0.027
PIP	FDP	−0.276	−0.281	0.011	0.400	−0.306	−0.007
	RB	−0.176	0.142	0.176	0.100	0.092	0.256
	UB	−0.176	0.141	−0.215	0.100	0.088	−0.238
	FDS	−0.276	−0.218	0.048	0.400	−0.263	0.029
	ES	−0.026	0.221	0.004	0.000	0.204	0.023
MP	FDP	−0.204	−0.302	0.022	0.300	−0.509	0.053
	FDS	−0.204	−0.352	0.035	0.300	−0.567	0.057
	RI	−0.404	−0.035	0.284	0.400	−0.302	0.316
	LU	−0.404	−0.112	0.186	0.400	−0.477	0.265
	UI	−0.404	0.047	−0.197	0.400	−0.240	−0.244
	LE	−0.104	0.313	0.062	0.000	0.352	0.052

Table 1-2e   Tendon locations in little finger (mean value of 15 specimens).

Joint	Tendon	Distal Point			Proximal Point		
		X	Y	Z	X	Y	Z
IP	TE	0.010	0.196	−0.077	0.000	0.193	−0.079
	FDP	0.010	−0.232	−0.041	0.300	−0.238	−0.003
PIP	FDP	−0.196	−0.333	0.040	0.400	−0.364	−0.002
	RB	−0.096	0.198	0.220	0.100	0.131	0.253
	UB	−0.096	0.093	−0.317	0.100	0.120	−0.298
	FDS	−0.196	−0.268	0.050	0.400	−0.345	−0.008
	ES	0.054	0.259	−0.010	0.000	0.254	0.014
MP	FDP	−0.044	−0.427	0.080	0.300	−0.628	0.170
	FDS	−0.044	−0.508	0.111	0.300	−0.708	0.170
	RI	−0.244	0.027	0.418	0.400	−0.182	0.497
	LU	−0.244	−0.169	0.364	0.400	−0.553	0.469
	ADQ	−0.244	−0.023	−0.486	0.400	−0.309	−0.625
	LE	0.056	0.154	0.061	0.000	0.190	0.133

Table 1-3 Sign convention used for anatomical variation in fingers.

	Component	Sign Force or moment direction
Constraint force variation	Axial	+ Axial compression
	Dorsovolar	+ Dorsal subluxation − Volar subluxation
	Radioulnar	+ Radial dislocation − Ulnar dislocation
Moment arm variation	Axial	+ Radial twist (pronation) − Ulnar twist (supination)
	Radioulnar	+ Radial bending − Ulnar bending
	Flexion-extension	+ Flexion − Extension

The normalized force potential and moment potential parameters of the hand model in neutral position are tabulated in Tables 1-4a to 1-4e. The sign convention used for the force potential and moment potential is illustrated in Table 1-3.

Table 1-4a Force and Moment Potential Parameters of Tendons in Normal Thumb under Neutral Position (mean value of 5 specimens).

Joint	Tendon	Force			Moment		
		ϕ	β	γ	a	b	c
IP	EPL	0.986	0.046	0.066	0.019	−0.041	−0.203
	FPL	0.954	−0.257	−0.039	0.025	0.037	0.281
MP	EPL	0.946	−0.111	−0.219	−0.083	−0.228	−0.233
	FPL	0.892	−0.363	−0.037	0.021	−0.007	0.445
	ADD	0.802	−0.562	−0.106	−0.141	−0.271	0.411
	EPB	0.950	0.012	−0.251	−0.078	−0.031	−0.260
	FPB	0.747	−0.481	0.224	0.126	0.330	0.287
	APB	0.807	−0.382	0.394	0.162	0.400	0.093
CMC	EPL	0.606	0.098	0.663	0.242	−0.014	−0.156
	FPL	0.707	0.458	−0.324	0.226	−0.180	0.170
	ADD	0.502	0.336	−0.195	0.256	−0.435	−0.160
	EPB	0.771	−0.321	0.279	0.148	0.131	−0.143
	FPB	0.493	0.350	−0.556	0.273	−0.441	0.040
	APB	0.484	−0.128	−0.666	0.250	0.318	0.143
	OPP	0.801	−0.411	−0.263	0.112	−0.035	0.377
	APL	0.851	−0.358	0.046	0.097	0.215	−0.164

Table 1-4b   Force and Moment Potential Parameters of Tendons in Normal Index Finger under Neutral Position (mean value of 15 specimens).

Joint	Tendon	Force			Moment		
		α	β	γ	a	b	c
DIP	TE	1.000	−0.015	0.005	0.001	−0.009	−0.196
	FDP	0.992	−0.115	0.053	−0.011	0.038	0.208
PIP	FDP	0.994	−0.112	0.019	−0.003	0.019	0.362
	RB	0.996	−0.010	0.090	0.039	0.258	−0.181
	UB	0.998	−0.040	−0.045	−0.017	−0.307	−0.135
	FDS	0.998	−0.069	0.014	0.005	0.022	0.283
	ES	0.999	−0.048	0.002	0.000	−0.026	−0.266
MP	FDP	0.964	−0.264	−0.030	0.031	0.013	0.517
	FDS	0.969	−0.242	−0.046	0.010	−0.097	0.595
	RI	0.950	−0.271	0.154	0.143	0.536	0.235
	LU	0.892	−0.431	0.133	0.153	0.430	0.456
	UI	0.959	−0.283	0.016	−0.113	−0.430	0.250
	LE	0.991	0.136	0.016	0.002	−0.026	−0.478

Table 1-4c   Force and Moment Potential Parameters of Tendons in Normal Middle Finger under Neutral Position (mean value of 15 specimens).

Joint	Tendon	Force			Moment		
		α	β	γ	a	b	c
DIP	TE	0.998	0.056	−0.014	−0.001	−0.017	−0.169
	FDP	0.988	−0.155	−0.008	0.009	0.052	0.190
PIP	FDP	0.998	0.063	−0.017	0.005	0.000	0.276
	RB	0.988	−0.146	0.049	0.044	0.254	−0.145
	UB	0.993	−0.106	−0.055	−0.023	−0.282	−0.089
	FDS	0.999	−0.035	−0.034	0.006	0.037	0.234
	ES	0.999	−0.027	0.042	0.005	−0.009	−0.234
MP	FDP	0.976	−0.216	−0.009	0.029	0.004	0.445
	FDS	0.963	−0.270	−0.002	0.013	0.061	0.490
	RI	0.907	−0.403	0.119	0.125	0.398	0.292
	LU	0.913	−0.402	0.073	0.137	0.339	0.460
	UI	0.986	−0.168	0.013	−0.045	−0.123	0.115
	LE	0.990	0.142	0.005	0.036	−0.013	−0.412

Table 1-4d Force and Moment Potential Parameters of Tendons in Normal Ring Finger under Neutral Position (mean value of 15 specimens).

Joint	Tendon	Force			Moment		
		α	β	γ	a	b	c
DIP	TE	0.998	0.060	−0.025	−0.003	−0.021	−0.154
	FDP	0.987	−0.150	0.050	−0.010	0.012	0.191
PIP	FDP	0.999	−0.028	−0.020	0.006	0.001	0.295
	RB	0.983	−0.098	0.156	0.039	0.236	−0.100
	UB	0.993	−0.105	−0.046	−0.026	−0.232	−0.098
	FDS	0.999	−0.050	−0.021	0.007	0.037	0.243
	ES	0.995	−0.067	0.075	0.026	0.023	−0.203
MP	FDP	0.971	−0.237	0.035	−0.006	0.041	0.423
	FDS	0.969	−0.246	0.025	−0.001	0.048	0.476
	RI	0.974	−0.226	0.028	0.063	0.297	0.204
	LU	0.951	−0.302	0.066	0.040	0.226	0.333
	UI	0.969	−0.242	−0.040	−0.053	−0.221	0.136
	LE	0.996	0.087	−0.023	−0.013	0.052	−0.351

Table 1-4e Force and Moment Potential Parameters of Tendons in Normal Little Finger under Neutral Position (mean value of 15 specimens).

Joint	Tendon	Force			Moment		
		α	β	γ	a	b	c
DIP	TE	1.000	−0.015	−0.010	−0.003	−0.079	−0.193
	FDP	0.997	−0.010	0.073	−0.013	−0.025	0.234
PIP	FDP	0.998	−0.035	−0.047	0.019	0.017	0.350
	RB	0.989	−0.132	0.067	0.049	0.244	−0.143
	UB	0.998	0.054	0.039	0.017	−0.301	−0.114
	FDS	0.994	−0.084	−0.064	0.009	0.018	0.309
	ES	0.995	−0.018	0.096	0.024	0.014	−0.253
MP	FDP	0.968	−0.229	0.103	−0.045	0.134	0.539
	FDS	0.971	−0.228	0.068	−0.033	0.145	0.619
	RI	0.982	−0.178	0.067	0.066	0.461	0.107
	LU	0.945	−0.316	0.086	0.126	0.409	0.396
	ADQ	0.964	−0.239	−0.116	−0.083	−0.556	0.202
	LE	0.984	0.079	0.158	0.043	0.131	−0.187

Representative tendon excursions of the seven muscles versus the flexion angle of the MP joint of the index finger are delineated (Fig. 1-7a). Excursions

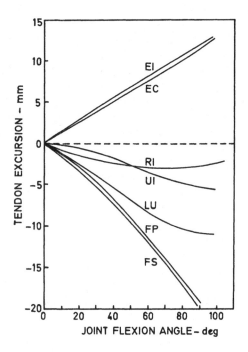

Fig. 1-7a   Tendon and intrinsic muscle excursion as a function of joint flexion angle.

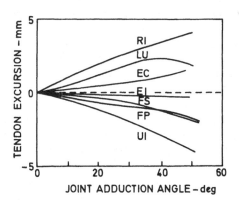

Fig. 1-7b   Tendon and intrinsic muscle excursion as a function of joint adduction (ulnar deviation) angle.

of the four extrinsic muscles were relatively large and almost linearly increased with the flexion angle. The flexors had larger excursions than the extensors. For intrinsic muscles, both interosseous muscles had relatively smaller excursions through the range of MP joint flexion, and most of their excursion occurred in the middle range of motion. However, the lumbrical muscle had a fairly large excursion which was almost equivalent to that of the extrinsic muscles.

Typical tendon excursions of the seven muscles during abduction and adduction motion of the MP joint of the index finger in an extended position were displayed (Fig. 1–7b).

In Table 1-5, averaged values of the tendon excursions, per 100 degrees of rotation of each joint in seven index finger specimens, are summarized. Averaged values and standard deviations (in parenthesis) of the associated joint range of motion are also included.

Table 1-5 Excursions of Index Finger Muscles*

Joint	Motion	Range of Motion, degree	Excursion (mm)/100 degrees of Joint Rotation						
			FP	FS	EI	EC	LI	RI	UI
DIP	Flexion/	50.36	7.30	—	2.94	3.04	1.12	—	2.39
	Extension	(6.60)	(1.17)		(0.88)	(0.78)	(0.87)		(0.72)
PIP	Flexion/	89.5	14.06	10.81	5.37	5.58	2.99	—	4.72
	Extension	(11.7)	(2.03)	(1.92)	(0.51)	(1.09)	(1.11)		(.88)
MCP	Flexion/	85.3	18.98	21.04	15.07	13.96	15.26	5.88	7.87
	Extension	(18.4)	(2.58)	(2.50)	(2.42)	(2.24)	(3.97)	(2.29)	(2.37)
MCP	Abduction/	50.4	2.92	3.99	3.48	3.35	8.10	11.46	10.17
	Adduction	(6.6)	(2.23)	(2.66)	(3.13)	(1.82)	(3.22)	(3.20)	(3.19)

* n = 7; mean (S.D.)

A representative moment arm throughout the range of motion at the MP joint of the index finger is depicted (Figs. 1-8a and 1-8b). Extension moment arms of the extensor indicis and extensor digitorum communis were fairly constant throughout the range of motion. The flexion moment arms of flexor digitorum profundus and flexor digitorum superficialis were slightly increased with increasing flexion angles. For the three intrinsic muscles, their flexion-extension momentum arms at the MP joint had a variable pattern and they tended to decrease when the flexion angle increased beyond 45 degrees.

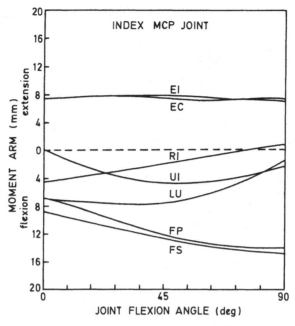

Fig. 1-8a   Tendon and intrinsic muscle moment arm at the MP joint as a function of flexion angle.

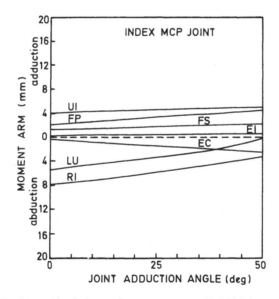

Fig. 1-8b   Tendon and intrinsic muscle moment arm at the MP joint as a function of adduction (ulnar deviation) angle.

For MP abduction-adduction, the three intrinsic muscles had consistently large moment arms which varied slightly throughout the range of motion (Fig. 1–8b). All of the extrinsic muscles, except extensor digitorum communis demonstrated adduction moment arms throughout the range of motion, although relatively small with values of 0.5 to 3.5 mm. The averaged values and standard deviation of the moment arms for the entire range of joint motion about the MP joint of the seven specimens studied are summarized in Table 1-6.

Table 1-6 Moment Arms of the Index Finger Muscles (averaged through the range of motion).

Joint	Motion	Muscle Moment Arms (mm)***						
		FP	FS	RI	LU	UI	EI	EC
DIP	Flexion/ Extension*	0.41 (0.14)	—	—	−0.07 (0.06)	−0.16 (0.05)	−0.19 (0.05)	−0.22 (0.04)
PIP	Flexion/ Extension*	0.79 (0.11)	0.62 (0.10)	—	−0.18 (0.13)	−0.26 (0.08)	−0.26 (0.11)	−0.28 (0.11)
MCP	Flexion/ Extension*	1.11 (0.11)	1.19 (0.07)	0.37 (0.14)	0.93 (0.21)	0.66 (0.21)	−0.90 (0.13)	−0.86 (0.16)
MCP	Abduction/ Adduction**	0.11 (0.17)	0.17 (0.20)	−0.61 (0.21)	−0.48 (0.16)	0.58 (0.17)	0.13 (0.16)	−0.02 (0.25)

*Flexion = + ; Extension = −
**Adduction = + ; Abduction = −
***Mean (S.D.) ; n = 7

DISCUSSION

Tendon locations and orientations in the finger were obtained by experimental studies of 15 normal hand specimens. The results presented are the average of these raw data. The measurements of tendon locations were normalized with respect to the middle phalanx length (O_2, O_3) of the corresponding finger before averaging. This normalization concept was based on the assumption that normal hands maintain similar anatomical structure and dimensional proportion, regardless of their physical size. With this normalization, anthropometric variation was minimized.

The results show small variations in predominant components, while significant variation occurred in secondary or minor components among the specimens studied. This indicates that tendons and muscles have more than one function. In addition to their predominant functions, each muscle also has other joint motion potentials. For example, in addition to flexion and extension, finger flexor and extensor tendons act as adductor of abductor muscles for all the fingers at the MP joints. However, these secondary contributions of function vary among the specimen population.

In addition to the radial and ulnar deviation moment, the intrinsic muscles also contribute to flexion at the MP joint. For all 4 fingers, the lumbricals have more flexion moment potential than the interossei. Varying amounts of moment potential for axial rotation were also noticed within each individual tendon and muscle. However, they were not as significant as those for flexion-extension and abduction-adduction.

For the purpose of verification, two-dimensional models, as proposed by Landsmeer[12], were used to compare with the results produced by the present model. Four parameters, namely x and y coordinates of the distal and proximal points, were used in the normative model to specify tendon locations.

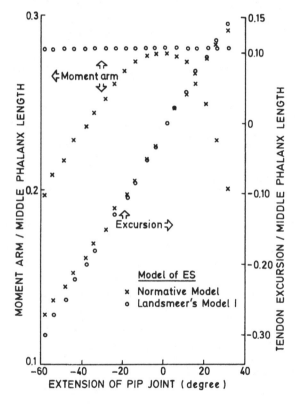

Fig. 1-9 Comparison of tendon excursions and moment arms of extensor slip at PIP joint obtained from the normative model with those of Landsmeer's Model I (in the range of 60° flexion to 30° extension).

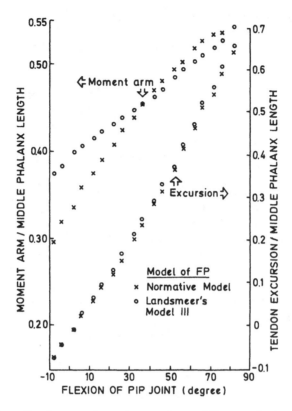

Fig. 1-10 Comparison of tendon excursion and moment arms of flexor profundus at PIP joint from the normative model with those of Landsmeer's Model III (in the range of 90° flexion to 10° extension).

With selection of various combinations of these four parameters, the results obtained could be approximately either similar to, or different from, those of Landsmeer's models. Figure 1-9 illustrates the comparison of tendon excursions and moment arms of the extensor slip at the PIP joint obtained from the normative model with those of Landsmeer's Model I. In the range of 60° flexion to 30° extension, the tendon excursions were almost the same. However, the moment arms were not quite the same. Those obtained from the normative model were smaller with the joint at more than 20° flexion or extension. This implies that the use of a bowstring model between two fixed points to represent certain extensor tendons may not be suitable. The proximal point of the bowstring model for the extensors should be allowed to shift along the articular surface under different joint rotation angles. The comparison of tendon excursions and moment arms of the flexor profundus tendon at the PIP joint obtained from the normative model with those of Landsmeer's Model III is shown in Fig. 1-10. The tendon excursions obtained from both models are similar, in the range of 90° flexion to 10° hyper-extension. The moment arms obtained from the normative model were smaller as the joint went into extension or under small flexion angle; however, they became larger at larger flexion angles. On the whole, the tendon excursions obtained by the normative model were frequently smaller than those from Landsmeer's model. Consequently, the joint and muscle forces estimated, based on the normative model, will be somewhat over-estimated when compared to those of Landsmeer's model.

Finally, at the PIP joint the conjoined lateral band shifted at the dorsolateral sides of the base of the middle phalanx[9,15]. When both interphalangeal joints flexed, the conjoined lateral band shifted several millimeters volarly. When the PIP and DIP joints extend, the lateral bands shift dorsally. This volar-dorsal tendon shift at the base of the middle phalanx, as shown in Fig. 1-11, was perfectly simulated by the two-point technique of the normative model. The present normative model can be used for both motion and force analysis of the normal hand. Systematic variation of these data based on clinical examinations can provide a simulated study of the hand in pathological conditions.

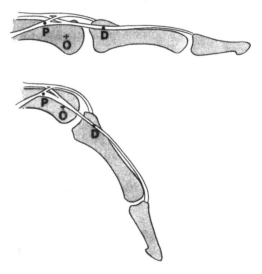

Fig. 1-11 Volar-dorsal shift of conjoined lateral band at PIP joint was illustrated by two-point technique of normative model.

REFERENCES

1. An, K.N., Chao, E.Y., Cooney, W.P. III and Linscheid, R.L., "Normative model of human hand for biomechanical analysis", *J. Biomechanics*, **12**:775-788 (1979).
2. Bunnell, S., *Surgery of the Hand*, 2nd edn, (Lippincott, Philadelphia, 1948).
3. Chao, E.Y., Opgrande, J.D. and Axmear, F.E., "Three-dimensional force analysis of finger joints in selected isometric hand functions", *J. Biomechanics*, **9**:387-396 (1976).
4. Doyle, J.R. and Blythe, W.F., "The finger flexor tendon sheath and pulleys: anatomy and reconstruction", *AAOS Symposium on Tendon Surgery in the Hand*, pp. 81-87 (1975).
5. Doyle, J.R. and Blythe, W.F., "Anatomy of the flexor tendon sheath and pulleys of the thumb", *J. Hand Surgery*, **2**:149-151 (1977).
6. Eyler, D.L. and Markee, J.E., "The anatomy and function of the intrinsic musculature of the fingers", *J. Bone Joint Surg.*, **36-A**:1-9 (1954).
7. Flatt, A.E., "The pathomechanics of ulnar drift", *Social and Rehabilitation Services, Final Report,* Grant No. RD 2226M (1971).
8. Haines, R.W., "The extensor apparatus of the finger", *J. Anat.*, **85**:251 (1951).
9. Harris, C. and Rutledge, A.L., "The functional anatomy of the extensor mechanism of finger", *J. Bone Joint Surg.*, **57-A**:713-726 (1972).
10. Landsmeer, J.M.F., "The anatomy of the dorsal aponeurosis of the human finger and its functional significance", *Anat. Rec.*, **104**:31 (1949).
11. Landsmeer, J.M.F., "Anatomical and functional investigations on the articulation of human fingers", *Acta Anat. Suppl.*, **24** (1955).
12. Landsmeer, J.M.F., "Study in the anatomy of articulation. The equilibrium of the 'intercalated' bone", *Acta Morph. Neerl.-Scand.*, **3**:287-303 (1961).
13. Shewsbury, M.M. and Kuczynski, K., "Flexor digitorum superficialis tendon in the fingers of the human hand", *The Hand*, **6**:121-133 (1974).
14. Smith, E.M., Juvinall, R.C., Bender, L.F. and Pearson, J.R., "Role of the finger flexors in rheumatoid deformities of the metacarpophalangeal joints", *Arthritis Rheum.* **7**:467-480 (1964).
15. Smith, R.J., "Balance and kinetics of the finger under normal and pathological conditions", *Clin. Orthop. Rel. Res.*, **104**:92-111 (1974).
16. Stack, H.G., "Muscle function in the finger", *J. Bone Joint Surg.*, **44-B**:899 (1962).

17. Zwieten, K.J. Van, "The extensor assembly of the finger in man and non-human primates", (Ph.D. Thesis, Univ. of Leiden, 1980.)
18. Brand, P.W., Beach, R.B. and Thompson, D.E., "Relative tension and potential excursion of muscles in the forearm and hand", *J. Hand Surg.*, **6**:209-219 (1981).
19. Storace, A. and Wolf, B., "Functional analysis of the role of finger tendons", *J. Biomechanics*, **12**:575-578 (1979).
20. Wood, G.A. and Jennings, L.S., "On the use of spline functions for data smoothing", *J. Biomechanics*, **12**:477-479 (1979).

2

QUANTITATIVE ANALYSIS OF THE INTRINSIC AND EXTRINSIC MUSCULATURE OF THE HAND

INTRODUCTION

The objective of this study was to provide a quantitative analysis of the intrinsic and extrinsic musculature of the hand on which a detailed three-dimensional biomechanical model could be based. Accurate determination of the volume and mean fiber length of the muscles allows determination of a physiologic cross-section from which an upper limit to the amount of force available for a specific muscle mass in a specific joint location can be approximated. By means of the techniques used in this study, the centroid of the muscle could be determined and used to estimate its resultant force vector in a theoretical analysis. These studies included the forearm musculature, allowing correlation between the muscle forces available to the extrinsic and intrinsic motors of the fingers. Moment arm values through which the muscle force acts were primarily determined through other studies.

Quantitative studies on the arm and forearm musculature are available from the studies of Fick[22] and others. Determination of the functional potential of the intrinsic muscles and their correlation with the forearm musculature has recently been reported by Brand[21]. Ketchum[15] measured individual intrinsic forces about the index metacarpophalangeal joint and correlated this with volumes of cadaver intrinsics. In this study, we attempted to include the contributions of individual components of the intrinsics where these were recognizable in order to distribute the force vectors as accurately as possible on a model.

FUNCTIONAL ANATOMY OF HAND MUSCULATURE

Qualitative and descriptive anatomy of the hand has been available from the time of Galen. Vesalius rekindled anatomic inquiry, thus spurring the incremental additions of this knowledge that have occurred sporadically in the intervening 400 years. The intrinsic muscles of the hand are commonly divided into the thenar musculature, the interosseous and lumbrical muscles of the four fingers, and the hypothenar muscles to the little finger. The thenar muscles consist of the abductor pollicis brevis, the opponens pollicis, and the flexor pollicis brevis. The former originates on the ridge of the trapezium, the tuberosity of the scaphoid, and the transverse carpal ligament. Classically, it is described as inserting on the radial capsule, the tubercle of the proximal phalanx of the thumb, and the radial aspect of the extensor apparatus. The opponens muscle lies just deep to this and originates from the trapezium and transverse carpal ligament and inserts along the distal two-thirds of the first metacarpal shaft. The flexor pollicis brevis arises deep to these muscles from two areas; the superficial from the edge of the transverse carpal ligament and the ridge of the trapezium and the deep head from the trapezoid. The muscle partially envelops the flexor pollicis longus tendon before inserting on the radial sesamoid, radial tubercle and occasionally the extensor expansion on the radial aspect of the thumb. This group of muscles acts in concert to provide opposition of the thumb. That is, a complex motion which consists of abduction of the thumb from the plane of the palm; flexion of the metacarpotrapezial joint; and abduction, flexion and pronation of the proximal phalanx on the metacarpal head for the purposes of positioning the thumb effectively against the pulps of the fingers for pinch and grasp functions.

Deep to these muscles lies the adductor pollicis, a complex fan-shaped muscle arising from two or three heads along the volar ridge of the third metacarpal and the radiovolar aspect of the capitate and trapezoid. The transverse portion of this muscle, arising from the distal half of the third metacarpal, passes virtually perpendicular to the third metacarpal to insert on the ulnar tubercle of the base of the proximal phalanx of the thumb. The oblique portion of the muscle, arising from the proximal portion of the third metacarpal and from the capitate, is separated from the transverse portion by a small hiatus created by the passage of the deep palmar arch and motor branch of the ulnar nerve. The fibers run distally and with slight rotation to the insert into the capsule and ulnar sesamoid. An accessory adductor muscle, when present, arises from the trapezoid base of the first metacarpal and inserts on the ulnar wing tendon of the extensor hood and is described by some as the first palmar interosseous. There is a fibrous tether to this tendon from the main adductor tendon.

The interosseous muscles of the hand are generally divided into dorsal and volar interossei with the former consisting of two heads from contiguous metacarpals. These converge in a bipennate fashion to insert into the tubercle of the base of the proximal phalanx on the side which will abduct it from the third metacarpal. A contribution to the extensor expansion on the same side varies with the particular interosseous and is considered, by some, a volar component of the dorsal interosseous. The palmar interossei have classically been considered three or four in number, depending on whether the accessory oblique head of the abductor pollicis is designated as first palmar interosseous. For reasons to be seen, we will not adopt this convention, though it is very appealing from the phylogenetic viewpoint to view the muscular components inserting into the extensor expansions in each location as palmar interossei. The named palmar interossei run from the deep aspect of the ulnar concavity of the second metacarpal and the radiovolar facet of the fourth and fifth metacarpals to insert exclusively on the extensor expansion and provide adduction toward the third metacarpal for the index, ring and little fingers. They also extend the interphalangeal joints.

The hypothenar muscles consist of the abductor digiti quinti, flexor digiti quinti and opponens digiti quinti. The former consists of a superficial portion which arises from the pisiform bone and adjacent pisohamate ligament, runs a ribbon-like course to the ulnar extensor expansion of the little finger. The deep portion arises from the ulnar aspect of the fifth metacarpal and inserts into the ulnar tubercle of the base of the proximal phalanx. The flexor digiti quinti brevis arises from the pisohamate ligament and inserts on the volar capsule. The opponens digiti quinti arises from two heads on the pisohamate ligament and hook of the hamate, passing disto-ulnarly to the volar ridge of the fifth metacarpal. The two heads are divided by the passage of the deep motor branch of the ulnar nerve and its accompanying artery. The hypothenar musculature helps to cup the hand for better spatial coherence to grasp objects by flexing and supinating the fifth metacarpal on the hamate and flexing the metacarpophalangeal (MCP) joint of the little finger. The fourth metacarpal is induced to follow the fifth by its ligamentous attachments largely provided through the transverse intermetacarpal ligaments.

The last group of muscles are the lumbricals which arise from the profundus tendons and insert usually into the wing tendons of the radial aspect of the extensor hoods. There are variations, particularly at the ring and little fingers.

The understanding of the functional anatomy of the hand was markedly enhanced by the interest in normal and pathological function generated by injuries to the hand in World War II, and subsequent interest in pathomechanics and reconstructive possibilities of the rheumatoid hand. The functions of the hand were divided into two broad classifications by Napier[16] as grasp and pinch with possible variations and mixtures of the two. Lister[17] has suggested eight basic hand functions which include: precision pinch, pulp pinch, key pinch, chuck grip, hook grip, span grip, power grasp and flat hand.

Pinch is that function in which the pulp of the thumb is juxtaposed to the side, tip or pulp of the adjacent fingers, usually the index or long. Such functional posture is generally carried out with the thumb and fingers in a concave arch and requires the stabilization of three joints in the finger, as well as the thumb. In the former, since the metacarpocarpal joint is rigidly fixed by its flat geometry and taut capsules, musculotendinous stabilization is required at the MCP and two interphalangeal (IP) joints. The thumb, because of the marked freedom of the metacarpotrapezial (MCT) joint, requires musculotendinous stabilization at its basilar joint, as well as its MCP and IP joints. Stabilization requires the balanced interaction of both the intrinsic and extrinsic muscles on the intervening joints so that an output force may be generated in the area of contact between thumb and fingers. In the thumb, pinch force is generated at the pulp by the flexor pollicis longus and requires stabilization of the MCP joint by the thenar muscles, primarily those components of the abductor pollicis brevis, the flexor pollicis brevis, and adductor pollicis which exert a flexion moment at the MCP joint. These muscles, together with the opponens pollicis and the contribution of the abductor pollicis longus, also help to stablilize the MCT joint.

When manipulation of the held object is required, subtle rhythmic variations in tensions of the various muscles allows the tip of the thumb and fingers to describe translational and rotational arcs through variations in the stance of the three involved joints.

The finger to resist pinch, as exemplified by the index, must balance the force applied by the thumb to retain static equilibrium. To do this requires a flexion stance of both IP joints and the MCP joint. Flexion moment is provided at the DIP and PIP joints by the flexor profundus and may be aided at the PIP joint by the flexor superficialis. The opposing thumb induces a pronation torque on the index finger. This is counteracted by the contribution of the first dorsal interosseous by virtue of its insertion into the radiovolar tubercle at the proximal phalanx. A supinating moment about the long axis of the finger is thus provided to resist the force from the thumb.

Manipulation of the finger requires variations of position of the three joints. Landsmeer[7] suggested that the unique contribution of the human hand is the ability to reciprocally angulate the MCP joint to the prevailing angulation of the joints. This allows for the fine, discrete motions of the thumb and fingers necessary for intricate tasks associated with human craftsmanship.

In the pinch configuration, therefore, the opposing thumb and fingers may be thought of biomechanically as 4-bar linkages which require not only active stabilization of the links upon one another during static configurations, but also during dynamic uses of the fingers. The extrinsic muscles spanning all four links during their excursion are better suited to provide a continuing output force which can be modulated by the fine adjustments taking place in the intrinsic musculature. The major angulatory adjustment occurring at the MCP joint is due to variation of the tensions in the nonarticular components of the dorsal interossei.

The intrinsic musculature adjusts IP position with wing tendon insertions of the dorsal interosseous, volar interossei and lumbricals. These mediate variations of tensions through the lateral bands on either side of the extensor apparatus which course dorsolaterally to the PIP joint and extend over the dorsum to the DIP joint. Motion in these structures in modulated by the extensor digitorum longus which adjusts the position of the extensor hood over the MCP joint, and thus acts as a sliding mechanical transducer. Interphalangeal joints in the normal state, function synchronously due to this unique configuration of the extensor apparatus on the dorsum of the finger. The matrix-like configuration of the fibers composing the extension apparatus and their contribution to the stable excursions of the IP joints have recently been well described by Von Zweiten *et al.*[24].

Grasp functions are carried out by placing the palm of the hand over the intended object. This generally requires extension of the wrist, fingers and thumb by the long extensors and intrinsic components inserting into the wing tendons. Grasp is carried out by initiating flexion of the fingers with the extrinsic flexors and continued by setting the grasp with the phalangeal inserting interossei.

Electrodiagnostic studies in the hand began with the keen observations of Duchenne[18] and have been carried further in recent years by Close[19], Basmajian[20], Long[9] and others. Axiomatically, those muscles in a suitable position to perform an intended motion adjacent to a joint become electrically active during the performance of the function. Antagonistic muscles modify or stabilize a joint either through their viscoelastic response or contractile lengthening. These studies are capable of showing small variations of function and are also capable of a semiquantitative analysis. Thus, functional electromyography has demonstrated a wide variation in electrical activity of the extensor digitorum communis. Marked activity occurs during phalangeal extension but minimal amounts of activity are frequently present, even during flexion motion. This tends to accord with the necessity of maintaining some tension on the extensor apparatus for proper positioning during all phases of phalangeal action. It may also represent the necessity of having a dynamic constraint force to aid the collateral ligaments of the MP joint when it is in the last 60 to 70° of flexion where the ligaments are relatively taut and under the greatest stress.

Lumbrical activity is noted only during IP extension, except for rare exceptions and is not a significant initiator of MCP flexion. It seems capable of acting as a mechanical transducer by transferring the visco-elastic tension in the profundus tendon to the extensor apparatus. Thomas *et al.*[25] ascribed a large mechanical advantage to this muscle, despite its small size which accrues from its palmar location and profundal origin which allows it to exert optimal tension at any position of the MCP joint. Superficialis activity is minimal during flexion and extension of the unresisted long finger, but may be called into play when the finger is flexing against resistance or when the wrist is flexed past neutral, thus modifying the visco-elastic tensile contribution of the superficialis to the proximal IP joint. Likewise, the interosseous, thenar and hypothenar muscles may be seen to become electrically active in accordance with the shape of the external object which is being manipulated. It is those muscles which are necessary to adapt a configuration consistent with grasping the object that was activated.

PHYLOGENETIC CONSIDERATIONS

The present anatomical structure of the human hand appears to have been modified by evolutionary changes from the primitive pentadactylic mammalian hand. This consisted of three layers of muscles, dorsal, volar and superficial, arranged as dorsal abductors, intermediate flexores breves, and superficial adductores (or contrahentes). During the evolutionary process, the dorsal muscles, which are phalangeal abductors from the third metacarpal ray, have retained their bipennate character as the four dorsal interossei. Phalangeal insertions of the abductor digiti quinti (ADQ) and either the flexor brevis (FPB) superficial head or abductor pollicis brevis (APB) may represent the border ray abductors. Only four palmar interossei retain their primitive character of the ten flexores breves of the intermediate muscle layer. These are on the ulnar aspect of the index finger, the radial aspect of the ring and little fingers, and ulnar border of the thumb (accessory head of the adductor pollicis). If the breves were numbered, as suggested by Cunningham[26], from #1 beginning on the radial aspect of the thumb through #10 on the ulnar aspect of the little finger, only #'s 2,4,7 and 9 have retained their original character. The accessory oblique head of the adductor pollicis probably represents the second flexor brevis. There is occasionally a small muscle inserting into the radial wing tendon of the index finger (M. interosseous prior indicis of Albinus) obviously separate from the first dorsal interosseous, representing the third flexor brevis. We found three examples of this muscle with the characteristics of a palmar interosseous in seven dissections. The third and fourth volar interossei represent the seventh and ninth flexores breves.

The second, third and fourth dorsal interossei (DIO_2, DIO_3, DIO_4) in the human finger represent the major problem in trying to accord with this theory. The third dorsal interosseous appears to insert entirely into the ulnar wing tendon. Migration and coalescence of the palmar interosseous into the dorsal interosseous of these three digits presents the composite muscle seen in these locations. Several authors have noted not only a separate volar component of the second, third and fourth dorsal interossei, but also three separate motor nerves for the innervation of the three muscular components. The second dorsal interosseous is usually rather easily resolved into the two dorsal components which insert into the proximal radial phalangeal tubercle and the palmar component with an insertion into the radial wing tendon. The fourth dorsal interosseous may be dissected similarly, but the palmar component is less easily identified. Landsmeer[17] described the convolutions of the muscle fibers and their tendons in successive cross sections of the fourth dorsal inter-osseous, DIO_4. This description makes it obvious that the complexity of the muscle prevents easy acceptance of a simplified dissection scheme. The third dorsal interosseous inserts primarily into the extensor expansion. Lewis[8] suggests that this may be explained by evolutionary recession of the expected phalangeal insertions to a new insertion onto the inner, or deep, aspect of the extensor expansion. The tenth flexor brevis is functionally represented by the superficial head of the abductor digiti quinti which inserts into the ulnar lateral band. Earlier morphologists felt that the tenth flexor brevis was represented by the deep portion of the opponens digiti quinti on the basis of its position to the ulnar nerve, a view refuted by Lewis[8].

If this model is used, the complexity of the insertions of the interossei is markedly clarified and assessment of function is enhanced. The dorsal inter-ossei insert beneath the transverse laminae into the phalanx, while the flexores breves insert into the lateral wing tendon after passing superficially to the trans-verse laminae. Occasionally, the insertions recede into the laminae. The per-centage of the dorsal interosseous muscles inserting into the proximal phalanx was averaged by Eyler and Markee[2] as first dorsal interosseous 98.4, second dorsal interosseous 68.6, third dorsal interosseous 5.6, and fourth dorsal inter-osseous 43.1. Similar findings were noted by Salisbury[27] and others. It is rea-sonable, perhaps, to consider the remaining percentage of these muscles, except for the third, to represent flexores breves components.

Of the volar interossei, the percentage of muscle mass insertion into the extensor apparatus was first volar 98.9, second volar 99.6 and third volar 94.2, suggesting the volar IC's are essentially unipennate muscles inserting almost exclusively into the wing tendons. The first and second lumbricals insert ex-clusively into the radial wing tendon, whereas an occasional variation was noted in the third and fourth lumbricals, though these two largely inserted into the wing tendons with infrequent bony insertions.

Eyler and Markee[2] also noted that volar interossei were roughly twice as great in mass as the corresponding lumbricals and half the corresponding dorsal interossei. While this provides a rough measure of the muscle volume available for the specific functions across the finger joints, it is not sufficiently accurate for three-dimensional biomechanical analysis. With this anatomic knowledge clearly understood, a quantitative hand muscle analysis could then be initiated.

METHODS AND MATERIALS

Eleven hand and forearm specimens were studied by using two different methods. In the first method, fresh frozen forearms from cadavers or amputations in patients with peripheral malignancy were used for cross-sectional studies. The hand and forearm were placed in a light custom-made plywood box in the extended position while still frozen. A foam resin elastomer (brand name Pedilen foam) was poured so that the forearm and hand were completely supported by the plastic. The box and arm were then refrozen for several days to ensure rigidity, and an X-ray of the box and contained arm was made.

A movie camera was focused at the center of the limb cross section and two perpendicular lines at the edge of the box were used to provide a coordinate system for later measurements. The hand was then sectioned at ½ cm intervals to the level of the radial metaphysis and then at 1 cm intervals through the elbow. After each cut, loose dust and ice was removed from the cross section. Several frames of film were exposed while altering the lens aperture. On one arm, the sections through the hand were taken at 2.5 mm intervals.

The developed film was projected on a VanGuard Motion Analyzer. The X-Y coordinate system on the box was locked into the analyzer coordinate system. The superimposed centimeter ruler on the side of the picture was used to provide a scale factor to obtain the actual dimensions. The Z-axis was parallel to the long axis of the forearm, perpendicular to the X-Y axes, and measurements along this axis were determined by the interval between cuts. Individual points were then taken around the perimeter, first of the metacarpals, then the individual intrinsic muscles. On the first four hands, this was done by using a punch card system obtaining approximately 20 points per muscle. Coordinates were usually obtained by moving the intercept of the analyzer from left to right and obtaining two values of Y for each value of X.

Certain complex configurations made it necessary to reverse the data point collection or to divide the structure into two separate divisions and analyze it separately. It was sometimes necessary to run the film backward and forward several times to ensure accuracy in tracing individual muscles. Both bones and muscles were traced from where the bone first entered the initial proximal cut to its termination. In the case of muscles, this was followed to the last section in which muscle fibers were still apparent on the tendon insertion. On the fifth hand, which was studied at 2.5 mm cuts, analysis was carried out using a Sonic Digitizer (brand name GRAF-Pen System) in which 25 points could be entered for each perimeter. The current method has been reported in several previous studies[24,28,29].

The data was then transcribed to magnetic tape for processing by computer. The area and centroid of each cross-sectional element was calculated. At the completion of processing the data for each structure, the cross sections were summed, allowing computation of the total volume and longitudinal trajectory of the muscle centroidal line. Because the centroids of the metacarpals were also plotted, a moment arm of the muscle in relation to the center of the MCP joint, as well as other bony locations, was also obtained. The individual sections were stored and dissected and additional still photographs made so that verification of the appearance of the section of the individual muscles could be obtained.

The cross-sectional cuts were performed perpendicular to the third metacarpal. Because the other metacarpals deviate from this, the cross section of muscles on adjacent metacarpals are at a small angle to their longitudinal axes. This does not affect the calculation of the volume, but the length of the muscle which is determined by the distance between sections will have a slight error with measured length less than actual length. This can be easily corrected should such variation become important.

We chose not to attempt to treat muscles with several components, i.e. the dorsal interossei, as consisting of two or more separate cross sections because of the technical difficulty of doing so. In the second, third and fourth dorsal interossei, the two heads have a very small divergent angle, and especially in the latter separation of the two heads from the volar component is quite difficult due to the spiral intertwining of the muscular fascicles. The importance of obtaining accurate cross sections perpendicular to muscle fibers was deemed minimal as due to vector addition, the force exerted along the centroidal line of the whole muscle will be proportional to a cross section perpendicular to this centroidal line rather than a cross section perpendicular to the muscular fascicles of each head.

Five additional hands were dissected by conventional techniques, carefully removing the muscle, weighing it and measuring its volume by immersion in a graduated cylinder. These hands were in a rest position, the fingers being stretched out to the normal position of rest before commencing dissection. The length of the individual intrinsic muscles, therefore, was assumed to be at, or near, the rest length and care was taken to avoid stretching during the dissection process. It became obvious that muscle length measured in this manner was invalid. During dissection, an attempt was made to follow the muscle from its insertion to its origin, thereby separating individual components which had either an extensor expansion insertion or a phalangeal insertion. Each component was then weighed, and recorded separately and drawings completed during the process of dissection.

Subsequently, we then dissected seven hands measuring volume of individual components. In addition, the mean fiber length for each muscle was measured following the method of Brand *et al.*[21]. The physiological cross-

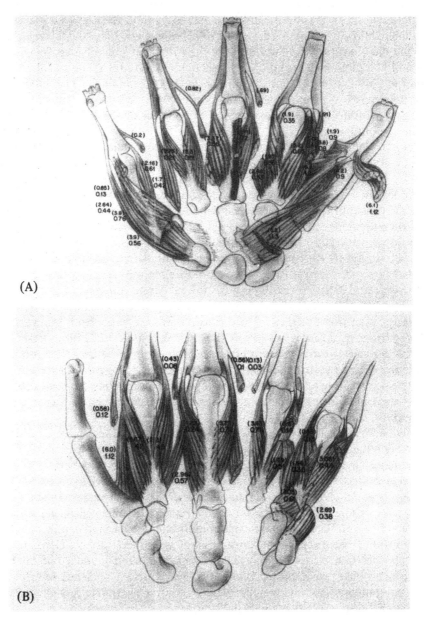

Fig. 2-1 The exploded view of the hand intrinsic muscles. The volume of each muscle is express-ed in the number enclosed in the parenthesis and the paired number represents the physiological cross-sectional area. (A)Volar view, (B)Dorsal view.

sectional area (PCSA) was defined as the volume of a muscle divided by its mean fiber length rather than the maximum measured cross section as defined by Fick[22]. These intrinsic muscle drawings were later composed to show an exploded view of the structural elements with their volume and PCSA illustrated (Fig. 2–1).

Because of our inability to separate the first palmar interosseous (accessory adductor pollicis oblique) in the cross-sectional studies and its absence or small size in the early dissected specimens, we elected to include this in the total mass of the adductor pollicis. Thus, we continued to number the volar interossei as numbers one through three. As will be seen in the tables of the later dissections, the accessory adductor is listed both under the adductor and the flexor breves. We have separately determined volume, fiber length of the volar muscular components of the dorsal interossei inserting into the extensor apparatus (except 3rd DIO) as flexor breves with the palmar interossei.

Moment arms at the MP joints were measured on the cross sections of three hand specimens. A point was established in the metacarpal head as a presumed center of rotation working back proximally from the closest cross section of the MCP joint contact surface. The radial distance to the center of each tendon, as well as the vertical (AP) moment arm Y and the lateral moment arm Z, were all determined. These values correlate closely with those made by using a biplanar radiographic technique (Chap. 1). These also appear to correlate well with those given by Landsmeer and Ketchum[7,15].

HAND MUSCLE MORPHOMETRICS

Table 2–1 provides a glossary of abbreviations used for all the intrinsic and extrinsic muscles studied. Tables 2–2 and 2–3 summarize the morphometric data for the intrinsic muscles of seven specimens and the extrinsic muscles of four specimens in which fiber lengths were measured. The intrinsic muscles are listed vertically beginning with the thenar muscles, dorsal interossei, palmar interossei, lumbricals and hypothenar muscles. The adductor pollicis (divided into three branches) consistently had the largest volume, followed by the opponens pollicis and the first dorsal interosseous. The lumbricals represent the smallest muscle volume for, particularly, the third and fourth lumbricals. The muscle PCSA has slightly different distribution due to the variation in mean fiber length. The percentage distribution for muscle volume and PCSA provides a better visualization of the contribution of different muscle groups in performing hand function.

The radial and ulnar components of the dorsal interossei show comparable volumes, except for the first DIO where the radial component originating on the first metacarpal is larger. This appears secondary to the necessity for

Table 2-1 A glossary of definitions and abbreviations used for the analysis of hand musculature

Abbreviation	Definition	Abbreviation	Definition
BR	Brachoradialis	FPL	Flexor Pollicis Longus
ECRL	Extensor Carpi Radialis Longus	FPB	Flexor Pollicis Brevis
		FP2	Flexor Profundus Index
ECRB	Extensor Carpi Radialis Brevis	FP3	Flexor Profundus Middle
		FP4	Flexor Profundus Ring
SUP	Supinator	FP5	Flexor Profundus Little
EDC2	Extensor Digitorum Communis Index	PQ	Pronator Quadratus
		APB	Abductor Pollicis Brevis
EDC3	Extensor Digitorum Communis Middle	OPP	Opponens Pollicis
		FPBsup	Flexor Pollicis Brevis Superficial Head
EDC4	Extensor Digitorum Communis Ring	FPBdeep	Flexor Pollicis Brevis Deep Head
EDC5	Extensor Digitorum Communis Little	ADPT	Adductor Pollicis Brevis Transverse Head
ECU	Extensor Carpi Ulnaris	ADP2	Adductor Pollicis Brevis Oblique Head 2nd MC
APL	Abductor Pollicis Longus	ADP3	Adductor Pollicis Brevis Oblique Head 3rd MC
EPB	Extensor Pollicis Brevis	DIO1R	Dorsal Interosseous 1st Radial Head
EPL	Extensor Pollicis Longus	DIO1U	Dorsal Interosseous 1st Ulnar Head
EIP	Extensor Indicis Proprius		
EDQ	Extensor Digitorum Quinti	ADPacc	Adductor Pollicis Accessory Head, (FB2)*
PT	Pronator Teres	DIO1F3	Dorsal Interosseous Superficial Head (FB3)*
FCR	Flexor Carpi Radialis	PIO1F4	Palmar Interosseous PIO1 (FB4)*
PL	Palmaris Longus		
FCU	Flexor Carpi Ulnaris	L2	Lumbrical Index
FSD	Flexor Superficialis Digastric Head	L3	Lumbrical Middle
		L4	Lumbrical Ring
FS2	Flexor Superficialis Index	L5	Lumbrical Little
		ADQS	Abductor Digiti Quinti Superficial Head
FS3	Flexor Superficialis Middle	ADQD	Abductor Digiti Quinti Deep Head
FS4	Flexor Superficialis Ring	FDQ	Flexor Digiti Quinti
FS5	Flexor Superficialis Little	OPPL	Opponens Digiti Quinti

*(FB) = flexoris brevis: a uni pennate muscle of the intermediate muscle layer that inserts into the extensor hood. It may be a palmar interosseous or a superficial head of a dorsal interosseous (see text).

Table 2-2 Intrinsic muscles of the hand (n = 7)

Muscle	Volume (cm³)	Fiber Length (cm)	PCSA (cm²)	% Volume	%PCSA
ABP	4.8 ± 1.8	3.2 ± 0.6	1.5 ± .6	6.2 ± 1.4	4.4 ± 1.0
OPP	6.1 ± 3.0	2.3 ± 0.6	2.8 ± 1.3	7.3 ± 1.7	7.5 ± 2.7
ADPT	3.4 ± 1.6	3.8 ± 0.6	0.9 ± 0.4	4.2 ± 1.4	2.5 ± 1.0
ADP2	2.1 ± 2.0	1.7 ± 1.6	0.6 ± 0.6	2.4 ± 1.9	1.6 ± 1.6
ADP3	6.1 ± 3.7	3.5 ± 0.8	1.8 ± 1.2	7.8 ± 3.5	5.2 ± 2.8
ADPacc	1.6 ± 0.6	1.8 ± 0.2	0.8 ± 0.3	2.0 ± 0.3	2.4 ± 0.6
FPBsup	3.9 ± 1.2	3.1 ± 1.0	1.3 ± 0.6	5.4 ± 2.4	4.1 ± 1.6
FPBdeep	2.1 ± 1.2	2.1 ± 0.3	1.0 ± 0.5	2.8 ± 1.4	3.0 ± 1.5
DIO1R	6.0 ± 2.0	2.8 ± 0.6	2.1 ± 0.5	7.8 ± 1.6	6.5 ± 2.3
DIO1U	3.5 ± 1.5	1.7 ± 0.5	2.0 ± 0.8	4.2 ± 1.4	5.5 ± 1.6
DIO2R	1.8 ± 1.1	1.2 ± 0.2	1.4 ± 1.0	2.0 ± 0.7	3.7 ± 1.8
DIO2U	1.9 ± 0.9	1.3 ± 0.2	1.4 ± 0.8	2.5 ± 1.0	4.4 ± 2.4
DIO3R	1.4 ± 0.9	1.3 ± 0.3	1.0 ± 0.8	1.6 ± 0.8	2.7 ± 1.4
DIO3U	1.6 ± 0.8	1.3 ± 0.3	1.2 ± 0.7	2.0 ± 0.4	3.6 ± 1.3
DIO4R	1.3 ± 1.0	1.1 ± 0.2	1.1 ± 0.8	1.5 ± 0.9	2.8 ± 1.4
DIO4U	1.4 ± 0.8	1.3 ± 0.3	1.1 ± 0.7	1.7 ± 0.5	2.9 ± 1.0
ADPF2*	1.6 ± 0.6	0.0 ± 0.0	0.0 ± 0.0	2.0 ± 0.3	0.0 ± 0.0
DIO1F3*	0.3 ± 0.5	0.7 ± 0.9	0.1 ± 0.2	0.3 ± 0.5	0.4 ± 0.7
PIO1F4*	2.5 ± 0.7	1.7 ± 0.2	1.4 ± 0.4	3.3 ± 0.7	4.1 ± 0.9
DIO2F5*	1.2 ± 0.7	1.6 ± 0.2	0.7 ± 0.4	1.5 ± 0.7	2.0 ± 0.9
DIO3F6*	1.1 ± 0.4	1.6 ± 0.3	0.7 ± 0.2	1.5 ± 0.4	2.0 ± 0.6
PIO2F7*	2.2 ± 0.7	1.7 ± 0.1	1.2 ± 0.3	2.8 ± 0.4	3.5 ± 0.7
DIO4F8*	1.0 ± 0.4	1.6 ± 0.2	0.6 ± 0.2	1.3 ± 0.4	1.9 ± 0.6
PIO3F9*	2.2 ± 0.7	1.6 ± 0.1	1.3 ± 0.4	2.9 ± 0.5	4.0 ± 1.1
L2	1.7 ± 0.7	4.7 ± 0.9	0.3 ± 0.1	2.1 ± 0.4	1.0 ± 0.0
L3	1.2 ± 0.4	5.0 ± 1.4	0.2 ± 0.0	1.6 ± 0.4	0.7 ± 0.2
L4	1.0 ± 0.1	4.6 ± 1.5	0.2 ± 0.0	1.4 ± 0.4	0.7 ± 0.3
L5	0.7 ± 0.3	3.5 ± 0.8	0.2 ± 0.0	0.9 ± 0.1	0.6 ± 0.1
ADQS	3.9 ± 2.0	4.3 ± 1.0	0.8 ± 0.3	4.8 ± 1.2	2.5 ± 0.5
ADQD	3.1 ± 1.2	3.4 ± 0.9	0.9 ± 0.2	3.9 ± 0.7	2.7 ± 0.7
FDQ	1.1 ± 1.0	7.7 ± 1.2	0.4 ± 0.4	1.5 ± 1.0	1.3 ± 1.0
OPPL	3.8 ± 1.4	1.3 ± 0.2	2.9 ± 1.1	5.0 ± 1.1	8.4 ± 2.6

* see footnote in Table 2-1

increased length of this component as the arc of motion of the thumb increased during hominid evolution. While the fiber length of the ulnar head is quite uniform, the fiber length of the radial head increases in a linear fashion, the further distal the origin on the first metacarpal.

The fiber lengths of the bipennate portions of the dorsal interossei are quite uniform at approximately 1.3 cm, while the fiber length of the volar components average nearly 1.6 cm and slightly less than the "true" palmar interossei. The increased fiber length would appear to represent the necessary added length to allow IP motion within their physiologic excurison.

Table 2-3 Extrinsic muscles of the hand (n = 4)

Muscle	Volume (cm³)	Fiber Length (cm)	PCSA (cm²)	% Volume	%PCSA
BR	46.9 ± 22.0	15.8 ± 5.6	2.9 ± 0.7	8.5 ± 1.5	2.9 ± 0.3
ECRL	32.4 ± 10.1	8.0 ± 1.0	4.0 ± 1.0	6.2 ± 0.9	4.1 ± 0.8
ECRB	29.2 ± 11.7	5.8 ± 1.0	4.9 ± 1.6	5.3 ± 0.2	4.9 ± 0.4
SUP	24.0 ± 11.6	3.5 ± 0.8	7.3 ± 4.4	4.3 ± 1.1	6.8 ± 2.4
EDC2	6.5 ± 1.4	6.1 ± 1.4	1.1 ± 0.3	1.3 ± 0.4	1.1 ± 0.2
EDC3	12.8 ± 10.1	6.8 ± 1.7	1.7 ± 0.8	2.3 ± 1.2	1.8 ± 0.8
EDC4	8.1 ± 2.0	6.8 ± 0.8	1.2 ± 0.4	1.7 ± 0.7	1.3 ± 0.4
EDC5	3.0 ± 2.0	3.8 ± 2.6	0.5 ± 0.4	0.7 ± 0.5	0.7 ± 0.6
ECU	18.1 ± 14.6	4.8 ± 1.9	3.5 ± 2.1	3.4 ± 2.1	3.9 ± 2.4
APL	14.8 ± 5.6	3.9 ± 1.2	3.9 ± 2.0	2.8 ± 0.7	4.0 ± 1.5
EPB	6.0 ± 3.8	4.2 ± 0.8	1.3 ± 0.7	1.0 ± 0.4	1.3 ± 0.5
EPL	10.5 ± 5.8	5.2 ± 1.1	1.9 ± 0.8	1.8 ± 0.5	1.9 ± 0.6
EIP	6.9 ± 3.6	4.9 ± 1.0	1.3 ± 0.6	1.3 ± 0.5	1.4 ± 0.6
EDQ	8.8 ± 3.9	6.2 ± 2.8	1.5 ± 0.9	1.6 ± 0.2	1.5 ± 0.5
PT	34.2 ± 14.7	5.3 ± 1.1	6.6 ± 2.7	6.3 ± 1.2	6.5 ± 1.3
FCR	27.3 ± 11.4	5.2 ± 0.7	5.2 ± 1.9	5.0 ± 0.5	6.0 ± 0.7
PL	6.8 ± 3.2	4.4 ± 0.9	1.5 ± 0.6	1.2 ± 0.2	1.2 ± 0.4
FCU	34.4 ± 13.6	3.3 ± 0.7	10.0 ± 3.0	6.2 ± 0.3	10.2 ± 1.0
FSD	10.0 ± 11.6	4.0 ± 0.5	2.4 ± 2.8	1.5 ± 1.7	2.0 ± 2.4
FS2	15.1 ± 8.1	4.2 ± 0.9	3.6 ± 2.1	2.8 ± 1.1	3.6 ± 1.6
FS3	29.5 ± 15.9	6.5 ± 1.0	4.2 ± 1.8	5.3 ± 1.8	4.2 ± 0.7
FS4	17.8 ± 10.4	6.8 ± 1.9	2.4 ± 1.2	2.9 ± 1.4	2.2 ± 0.7
FS5	9.3 ± 5.2	4.4 ± 0.6	2.1 ± 1.4	1.7 ± 0.7	2.2 ± 1.1
FPL	23.5 ± 10.4	4.6 ± 1.2	5.1 ± 2.6	4.2 ± 0.5	5.0 ± 1.7
FP2	27.6 ± 16.1	6.7 ± 0.7	4.1 ± 2.4	4.8 ± 1.7	3.8 ± 1.2
FP3	27.8 ± 11.0	6.7 ± 0.6	4.1 ± 1.4	5.2 ± 0.9	4.1 ± 0.7
FP4	24.8 ± 9.6	6.4 ± 0.7	3.7 ± 1.1	4.7 ± 1.0	3.9 ± 0.8
FP5	14.8 ± 5.8	5.9 ± 0.8	2.5 ± 0.9	3.0 ± 1.2	2.7 ± 1.1
PQ	10.0 ± 5.2	3.1 ± 0.9	3.5 ± 2.1	1.9 ± 0.8	3.6 ± 1.7

There were a number of variations of origins and insertions, particularly with the APB, FPB and ADP. Of the former, one hand appeared to have a pure phalangeal insertion, while the remainder had primarily an extensor expansion insertion. Another appeared to have two separate muscle bellies of the APB with distinct extensor insertion (superficial head) and phalangeal insertion (deep head). The FPB had two heads of similar size in most specimens. The deep head usually inserted on the sesamoid and volar capsule. The superficial head on two occasions inserted directly into the tubercle of the proximal phalanx and in another appeared to insert into the extensor expansion. In the other four specimens, the superficial head appeared to blend into the radial aspect of the capsule. On five of these specimens, there was a separate head of origin of the oblique head of the ADP from the second metacarpal with a commensurately shorter fiber length. By the same token, there were variations in the hypothenar dissection, most notably absence of a FDQ and on another occasion a doubled proximal phalangeal insertion deep head of the ADQ.

The volar components of the second and fourth dorsal interossei were somewhat easier to separate than that of the third. This was done under low power magnification, but friability of the muscle tissue, small size, and the twisting of the fibers made this dissection difficult.

These results show the thenar musculature accounts for 48.3% of the volume and 42% of the physiologic cross section of the available intrinsic muscles. Of the dorsal interossei, the first is nearly twice as strong as the second dorsal interosseous and represents 44% of the total volume and 39% of the PCSA of the dorsal interossei. The third dorsal interosseous is 76% of the second in volume, but the fourth averages 97% of the third. The palmar interossei are significantly smaller than the dorsal, averaging approximately 70% of the three ulnar dorsal interossei. The lumbricals are about one-third as large as the volar interossei. On the hypothenar side of the hand, the abductor digiti quinti and opponens represent 17% of the volume and 15% of the PCSA of the total intrinsic musculature.

Among the extrinsics, the BR has the largest muscle volume, but since its mean fiber length is also large, it results in a rather low PCSA value. The FCU has the largest PCSA, followed by SUP and PT. When the muscle PCSA for the digital flexors are compared with that of the extensors, a ratio of 3.5 to 1 is found, which reflects the much stronger flexor power for the fingers and the thumb as opposed to the extension power. However, finger flexors may increase their contraction efficiency by extending the wrist. This helps to explain why, under forceful grip action, the wrist assumes a natural extended posture.

JOINT FUNCTION POTENTIAL

The product of muscle PCSA with its moment arm at a particular hand joint is defined as the joint functional potential. This parameter helps to assess the contribution of all muscles at a joint in producing flexion-extension and radio-ulnar deviation power. This definition has a different meaning than Fick's muscle work capacity, since he considered the muscle's maximum cross-sectional area and excursion length[22]. These data are expected to be useful in selecting the proper muscle for tendon transfer, to assess functional strength reduction under neuropathic conditions, and other pathologic changes in the hand that may involve muscle function.

Based on 15 normal hand specimens, the tendon moment arms at each joint for the index, long, ring and little fingers were determined. An additional five hands were studied for the thumb tendon moment arms. A biplanar radiographic method was used. Since these specimens did not have an intact forearm after the radiographic studies, they became inadequate for cross-sectional study. The PCSA information on the related muscles was obtained from a different group of specimens. It compares favourably with published data[21]. The MP joint functional or moment potential for the entire hand is summarized in a polar diagram, as shown in Fig. 2-2. The length of the vector represents the relative magnitude of moment potential for each muscle controlling the hand.

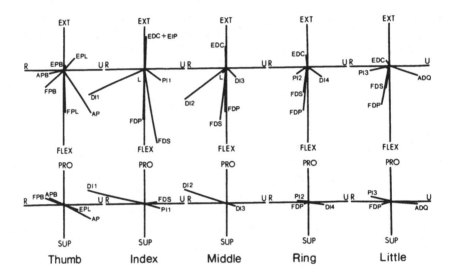

Fig. 2–2   Relative moment potential of the hand muscles around the MP joint for the fingers and the thumb.

The ratio of functional potential between the flexors and the extensors at the MP joints is higher than that predicted based on muscle PCSA alone. This is due to the added difference in tendon moment arm. To further illustrate these characteristics, seven additional specimens were studied for both the muscle PCSA and moment arm, and the results for the index finger MP joint are plotted (Fig. 2-3). Only the major flexors and extensors are included. The tendon excursion method was used to determine its moment (see Chap. 1). The ratio between flexion and extension moment potential is 5.4 to 1. However,

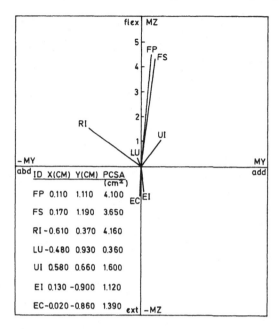

ID	X(CM)	Y(CM)	PCSA (cm²)
FP	0.110	1.110	4.100
FS	0.170	1.190	3.650
RI	-0.610	0.370	4.160
LU	-0.480	0.930	0.360
UI	0.580	0.660	1.600
EI	0.130	-0.900	1.120
EC	-0.020	-0.860	1.390

Fig. 2-3 Muscle moment potential for the index finger at the MCP joint.

these muscles seem to have slightly stronger radial deviation moment than ulnar deviation (2.7 units versus 2.1, respectively). This is primarily due to the radial interosseous muscle and its longer more efficient moment arm.

DISCUSSION

It is to be anticipated that the relative volumes, strengths and excursions of the intrinsic muscles will correlate with their functional demands. A study of these correlations provides some additional insight into the anatomic pre-requisites of hand function.

The importance of the thumb relative to the four opposing fingers seems to be reflected in the relative percentage of thumb muscle mass (TMM), 48% to the total intrinsic muscle mass (IMM) of the hand. The volume of muscle accorded to spanning each of the three joints of the thumb seems to accord well with the relative force necessary to stabilize each link on the cantilevered structure.

The opponens pollicis (OP), with 6.3% of the intrinsic mass, palmarly abducts, rotates and flexes the first metacarpal. The abductor pollicis brevis (APB), with nearly 7.3% of the total mass, spans the metacarpotrapezial (MCT) and metacarpophalangeal (MCP) joints. It will, thus, induce the same angulation of the first metacarpal on the trapezium as the opponens, but with a larger moment arm provides a somewhat greater torque. The relative amount of torque developed at the joint will depend on the stability and position of the MCP joint. In addition, that component of the muscle inserting on the radial wing tendon induces an extension moment at the interphalangeal (IP) joint. The flexor pollicis brevis (FPB), with nearly 9% of the IMM, has a smaller palmar abduction torque at the MCT, but a greater flexor torque at the MCP than the APB because of its different origin and insertion. These three muscles, with 23% of the intrinsic muscle mass, are involved in the function of opposition. The flexor pollicis longus (FPL) accounts for 4.4% of the total muscle mass of the forearm and hand, while the OP, APB and FPB account for 2.72% of the total muscle mass. The FPL is primarily a flexor of the IP joint, as can be seen in patients with paralysis of the ulnar nerve where there is a thumb stance of IP flexion, MCP extension and CMC flexion (Froment's sign).

Additional stability is provided at the MCP joint by the adductor pollicis (ADP) with the largest single muscle in the hand, 15% of intrinsic muscle mass and 1.9% total muscle mass. The transverse component of the muscle, approximately 33%, inserts on the ulno-volar tubercle of the proximal phalanx and flexes the MP joint. The oblique head, approximately 50%, inserts on the capsule and ulnar sesamoid, adducts the first metacarpal, and stabilizes the MP joint. The remaining 17% represents the accessory adductor which inserts on the ulnar wing tendon. The combined effect of the adductor produces a supination torque of the thumb ray about the longitudinal axes.

During key pinch in the index finger, the ability to resist the imposed torque about both the longitudinal axis and vertical axis of the MP joint is provided mostly by the radial interosseous and contributing EDC. For the index finger, the first dorsal interosseous represents 12% of the IMM and 1.6% of the TMM with a PCSA of 10.7% and 4.4%, respectively.

At the little finger, the ADQ has 9.0% of the IMM and 5.3% of the PCSA, whereas the PIO_3 is inserted entirely into the radial wing tendon with a usually insignificant lumbrical. The ADQ has approximately a 33% wing tendon insertion with the remainder inserted into the ulno-volar tubercle, sesamoid and capsule. The ulnar deviating force of this muscle is probably partially compensated by the radial moment from the extrinsic flexors, which approach the joint at approximately a 15° angle and bowstring to a certain extent during flexion. It also helps explain the vulnerability to ulnar deviation deformity of this joint in rheumatoid arthritis.

It is obvious that on the border fingers, index and little, the intrinsic muscle mass on the outer borders is significantly greater than on the internal fingers. On the latter, the radial and ulnar balance is better. It would appear that the border fingers compensate for the bowstringing effect of the extrinsic tendons, which induce centrally directed moment towards the third metacarpal.

It is not, at this point, clear how differently the phalangeal insertions of the intrinsics behave functionally as compared with the wing tendon insertions. As the latter are multiple joint muscles, their function is more complex. We would assume these would be active in a contractile fashion during adduction of the rays to the third metacarpal and during MP flexion and IP extension. They may act in a decelerating fashion during finger flexion to balance the angulatory acceleration of the distal finger joints. Much more sophisticated functioning electromyographic (EMG) studies are needed to determine the characteristics of the intrinsic components, both in normal and abnormal hands. They possess sufficient viscoelastic response to act as passive dampers to imposed extrinsic attenuations as well.

The concept of using the thickest portion of a muscle in successive cross-sectional cuts of a limb was suggested by Fick[22] and named the physiologic cross section, which he suggested should have a value of approximately 3.4 kg/cm². Other values have been suggested, but this value seems reasonably accurate for most work. While this method of obtaining the PCSA is reasonably accurate, Brand[21] pointed out that for a muscle with a long origin, but short fiber length, e.g. FCU, the value determined in this fashion would be too low and vice versa. He suggested that rotating the tendon of insertion from its most proximal appearance to its termination about the tendon or bone of origin would show the fibers to lie in a muscular sheet and be of remarkably similar length. Interestingly, this observation was made by a brilliant 17th Century anatomist named Nicolas Steno, but largely forgotten until recently. Sir Charles Bell also has a muscle represented as a parallelogram in his great treatise on the hand, but did not develop the concept further.

If, as Brand suggested[21], this measured fiber length with the muscle in the resting position is divided into muscle volume, a more accurate value for PCSA or as a ratio of individual PCSA to total PCSA a "tension fraction" is obtained. While this length has physiologic implications in regards to tension in the Blix curve that is not within the scope of the present discussion. If in simulation of the hand in a loaded condition where joint angulation varied significantly from the usual resting position, a modification of the value of the deliverable force as a function of its position on the length-tension curve would warrant a separate study. For the present, the relative values of these components combined with accurate moment arm measurements appear to provide reasonably accurate values for mathematical simulation of the hand.

Our hand muscle volume, PCSA and their distribution in the total hand musculature compared favourably with that published by Brand, *et al.*[21], although he uses the terms "mass fraction" and "tension fraction" to represent the percentage distribution of muscle mass and PCSA. Adding the moment arm information determined either by the biplanar radiographic method or the tension excursion technique has been an outstanding advantage in assessing muscle functional potential in moving the hand joints. A similar concept was adopted by Ketchum *et al.*[5] for the wrist joint. The present extrinsic muscle data have been combined with the remaining forearm musculature to perform the same analysis[23]. The functional potential concept has many applications in the surgery and care of an impaired and diseased hand. Although Fick[22] was the first to introduce such a concept, the present data have been substantially refined to enhance its reliability and applicability. These data should have direct applicability to clinical usage.

REFERENCES

1. Backhouse, K.M. and Catton, W.I., "An experimental study of the functions of the lumbrical muscles in the human hand", *J. Anat.,* **88**: 133–141 (1954).
2. Eyler, D.L. and Markee, J.E., "The anatomy and function of the intrinsic musculature of the fingers", *J. Bone Joint Surg.,* **36-A**: 1–9 (1954).
3. Haines, R.W., "The extensor apparatus of the finger", *J. Anat.,* **185**: 251–259 (1951).
4. Harris, C. and Rutledge, G.L., "The functional anatomy of the extensor mechanism of the finger", *J. Bone Joint Surg.,* **54-A**: 713–726 (1972).
5. Kaplan, E.B., *Functional and Surgical Anatomy of the Hand,* (Lippincott, Philadelphia, 2nd edn., 337 p, 1965).
6. Landsmeer, J.M.F., "The anatomy of the dorsal aponeurosis of the human finger and its functional significance", *Anat. Rec.,* **104**: 31–44 (1949).
7. Landsmeer, J.M.F., *Atlas of Anatomy of the Hand,* (Churchill Livingston, Edinburgh, London, New York, 349 p, 1976).
8. Lewis, O.J., "The evolution of the Mm. interossei in the primate hand", *Anat. Rec.,* **153**: 275–288 (1965).
9. Long, C., Conrad, P.W., Hall, E.A. and Furler, S.L., "Intrinsic-extrinsic muscle control of the hand in power grip and precision handling", *J. Bone Joint Surg.,* **52-A**: 853–867 (1970).
10. Smith, R.J., "Balance and kinetics of the fingers under normal and pathological conditions", *Clin. Orthop. Rel. Res.,* **104** (1974).
11. Srinivasan, H., "Patterns of movement of totally intrinsic-minus fingers", *J. Bone Joint Surg.,* **58-A**: 777–785 (1976).
12. Stack, H.G., "Muscle function in the fingers", *J. Bone Joint Surg.,* **44-B**: 899–909 (1962).
13. Micks, J.E., Reswick, J.B. and Hoger, D.L., "The mechanism of the

intrinsic-minus finger: a biomechanical study", *J. Hand Surg.,* **3**: 333–341 (1978).

14. Armstrong, T.J. and Chaffin, D.B., "An investigation of the relationship between displacements of the finger and wrist joint and the extrinsic finger flexor tendon", *J. Biomech.,* **11**: 119–128 (1978).

15. Ketchum, L.D., Brand, P.W., Thompson, D. and Pocock, G.S., "The determination of moments for extension of the wrist generated by muscles of the forearm", *J. Hand Surg.,* **3**: 205–210 (1978).

16. Napier, J.R., "The prehensile movements of the human hand", *J. Bone Joint Surg.,* **38-B**: 902–913 (1956).

17. Lister, G.D., Kleinert, H.E., Kutz, J.E. and Atasoy, E., "Primary repair of flexor tendon followed by immediate controlled mobilization", *J. Hand Surg.,* **2**: 441 (1977).

18. Duchenne, G.B., *Physiology of Motion,* trans. and ed. by E.B. Kaplan, (Lippincott, Philadelphia, 1949).

19. Close, J.R. and Kidd, C.C., "The functions of the muscles of the thumb, the index and long fingers", *J. Bone Joint Surg.,* **51-A**: 1601-1620 (1969).

20. Basmajian, J.V., *Muscles Alive,* (Baltimore, William S. Wilkins, 1962).

21. Brand, P.W., Beach, R.B. and Thompson, D.E., "Relative tension and potential excursion of muscles in the forearm and hand", *J. Hand Surg.,* **6**: 209–219 (1981).

22. Fick, A., "Statische berachtung der muskalture des oberschenkels", *Z. Rationelle Med.,* **9**: 94–106 (1850).

23. Gross, R.M., *A Quantitative Analysis of Forearm Musculature,* (M.S. Thesis, Dept. of Orthopedic Surgery, Mayo Graduate School of Medicine, 1979).

24. Zwieten, K.J. Van, *The Extensor Assembly of the Finger in Man and Non-Human Primates.* (Ph. D. Thesis, Univ. of Leiden, 1980).

25. Thomas, D.H., Long, C.L. and Landsmeer, J.M.F., "Biomechanical considerations of lumbricalis behavior in the human finger", *J. Biomech.,* **1**: 107-115 (1968).

26. Cunningham, D.J., "The intrinsic muscles of the hand of the thylacine, cuscus and phascogale", *J. Anat. Lond.,* **12**: 434–444 (1878).

27. Salisbury, C.R., "The interosseous muscles of the hand", *J. Anat. Lond.,* **71**: 395–403.

28. Rab, G.T., Chao, E.Y. and Stauffer, R.N., "Muscle force analysis of the lumbar spine", *Orthop. Clin. No. Am.,* **8**: 193–199 (1977).

29. Hui, F.C., Chao, E.Y. and An, K.N., "Muscle and joint forces at the elbow during isometric lifting", *Trans. of the 24th Mtg.: Orthopaedic Research Society,* **3**: 167 (1978).

3

MUSCLE AND JOINT FORCES IN THE HAND

The human hand is a three-dimensional structure. The finger and thumb are considered as a linkage system of intercalated bony segments balanced by muscle and tendon forces and joint constraints. Through the mathematical model developed based on the anatomical structure and functions of the hand, these internal forces were determined. However, the problem formulated was an indeterminate one. This redundant problem was analyzed by systematic reduction and optimization techniques. The internal force distributions under various isometric functions and daily activities were obtained. The potential application of these results in prosthetic joint design and testing and in the surgical reconstruction of pathologic hands is discussed.

INTRODUCTION

Restoration of function to fingers suffering from trauma or arthritis necessitates complete knowledge of the magnitude of tendon and joint forces, the manner in which they are transmitted, and their correspondence to one another. Quantitative studies in the past have been, in general, restricted to two-dimensional models[1,2,3,4] of simplified anatomical structures. Recently, several three-dimensional analyses have been performed[5,6]. However, the redundant problem was usually overlooked or treated by gross assumptions.

In this chapter, the three-dimensional force analyses of the finger and thumb are presented. The anatomical model was formulated based on a mechanical linkage system balanced by two groups of forces. The first group is the passive constraint forces and moments which are contributed by the joint articulating surfaces, ligamentous capsules, and the volar plate structure of the joints. The other group of forces is the active tensions developed in tendons (extrinsics) and muscles (intrinsics) oriented across the joint. The basic anatomical structure of the musculotendinous complex of the finger is shown in Fig. 3–1. In such a complex anatomical structure, assumptions were necessary in order to reduce it to a workable mechanical model. Ten normal hand specimens were analyzed in detail in order to establish a normative model

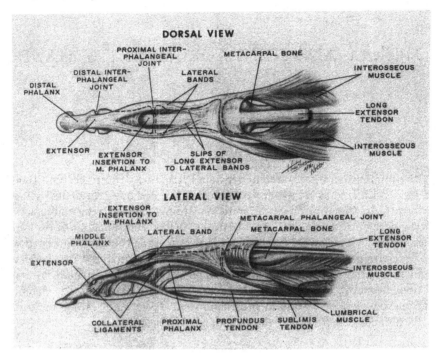

Fig. 3-1 Musculotendinous structure of a long finger.

describing the tendon locations and orientations in mathematical form. This model was presented in Chap. 1, and it will be used for the present analysis.

MATHEMATICAL MODELING OF THE HAND

According to the anatomical constraints and physiological functions of the fingers and thumb, different types of mechanical joints were assumed in modeling the system. The mechanical equivalent for each joint, its degrees of freedom, and the associated unknown constraint forces and moments are summarized in Table 3-1. Table 3-2 lists all the unknown tendons and intrinsic muscles which contribute to the isometric hand functions.

Six Cartesian coordinate systems at the index finger were defined, as shown in Fig. 3-2, to provide the orientation of the joints and the location of the tendons. The constraint forces (C_x, C_y, C_z) and moments (M_x, M_y) at the PIP joint are also shown. These forces and moments are the results of individual forces carried by the joint surface and ligamentous structures. The orientation angles between any two bony elements are described by a flexion-extension

Table 3-1 Equivalent mechanical model for the finger and thumb joints.

Hand Element	Joints	Mechanical Equivalent	Degrees of Freedom	Unknown Constraints
Finger	Distal interphalangeal joint (DIP)	Hinge joint	1	Axial compressive force (C_{x1}) Dorsovolar shear force (C_{y1}) Radio-ulnar shear force (C_{z1}) Axial twisting moment (M_{x1}) Radio-ulnar constraint moment (M_{y1})
	Proximal interphalangeal joint (PIP)	Hinge joint	1	Axial compressive force (C_{x3}) Dorsovolar shear force (C_{y3}) Radio-ulnar shear force (C_{z3}) Axial twisting moment (M_{x3}) Radio-ulnar constraint moment (M_{y3})
	Metacarpo-phalangeal joint (MP)	Universal joint	2	Axial compressive force (C_{x5}) Dorsovolar shear force (C_{y5}) Radio-ulnar shear force (C_{z5}) Axial twisting moment (M_{x5})
Thumb	Interphalangeal joint (IP)	Hinge joint	1	Axial compressive force (C_{x1}) Dorsovolar shear force (C_{y1}) Radio-ulnar shear force (C_{z1}) Axial twisting moment (M_{x1}) Radio-ulnar constraint moment (M_{y1})
	Metacarpo-phalangeal joint (MCP)	Universal joint	2	Axial compressive force (C_{x3}) Dorsovolar shear force (C_{y3}) Radio-ulnar shear force (C_{z3}) Axial twisting moment (M_{x3})
	Carpo-metacarpal joint (CMC)	Universal joint	2	Axial compressive force (C_{x5}) Dorsovolar shear force (C_{y5}) Radio-ulnar shear force (C_{z5}) Axial twisting moment (M_{x5})

angle (ϕ), radio-ulnar deviation angle (θ), and axial rotation (ψ). The rotational sequence of these angles, as related to both the distal (moving system) and proximal (fixed system) coordinate systems at a joint is illustrated in Fig. 3-3. In isometric force analysis, only joint orientation is important, and the motion pattern from one position to another can be ignored. Therefore, the noncommutational nature of the finite angular displacement will not be discussed here.

Table 3-2 Unknown tendon and intrinsic muscle forces involved in hand functions.

Hand Element	Joint	Unknown Tendon and Intrinsic Muscle Forces
Finger	DIP	Terminal extensor (TE) Flexor profundus (FP)
	PIP	Extensor slip (ES) Radial Band (RB) Ulnar band (UB) Flexor sublimis (FS)
	MP	Long extensor (LE) Radial interosseous (RI) Ulnar interosseous (UI) Lumbrical (LU)
Thumb	IP	Flexor pollicis longus (FPL) Extensor pollicis longus (EPL)
	MCP	Abductor pollicis brevis (APB) Flexor pollicis brevis (FPB) Adductor pollicis (ADD) Extensor pollicis brevis (EPB)
	CMC	Opponens pollicis (OPP) Abductor pollicis longus (APL)

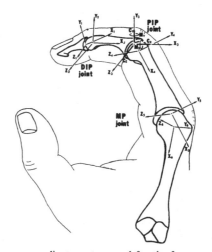

Fig. 3-2 Reference coordinate system used for the force analysis of finger.

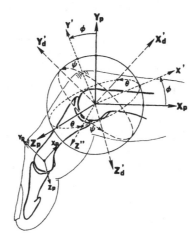

ϕ : Flexion-Extension (1st rotation — Z_p)
θ : Radioulnar Deviation (2nd rotation — Y')
ψ : Pronation-Supination (3rd rotation — X_d')

Fig. 3-3 Definition of finger orientation based on Eulerian angles.

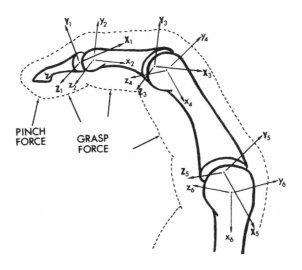

Fig. 3-4 Hypothetic externally applied forces in pinch and grasp.

Free-body analysis at each joint was then performed. The tensile forces carried by tendons or intrinsic muscles at each associated joint were assumed to be along these anatomical elements. The exact location and orientation of the tendon or muscle with respect to each joint were carefully defined, based on X-ray projection of finger specimens with metal markers embedded in the soft tissue. The detailed procedures of this technique and the results of a normative model are presented in Chap. 1 and in our previous publications[7-12]. The externally applied loads were assumed to be concentrated forces exerted at the tip of a finger or thumb during pinch or applied normally at the middle point of each plalanx during grasp, as shown in Fig. 3–4.

The equilibrium equations for the finger or thumb joints in performing isometric hand functions can be written in the following vector forms.

$$\sum_{i=1}^{n} |\bar{T}_i|\bar{e}_i + \bar{F} + \sum_{j=1}^{m} |\bar{A}_j|\bar{e}_j = 0$$

$$\sum_{i=1}^{n} |\bar{T}_i|(\bar{r}_i \times \bar{e}_i) + \bar{M} + \sum_{j=1}^{m} |\bar{A}_j|(\bar{S}_j \times \bar{e}_j) = 0$$

(1)

in which

$\bar{T}_i$ = tendon force vector,

$\bar{e}_i$ = unit vector in the direction of $\bar{T}_i$,

$\bar{F}$ = joint constraint force vector,

$\bar{A}_j$ = externally applied forces,

$\bar{e}_j$ = unit vector along $\bar{A}_j$,

$\bar{r}_i$ = position vector to $\bar{T}_i$,

$\bar{M}$ = constraint moment vector, and

$\bar{S}_j$ = position vector of $\bar{A}_j$.

Based on anatomic structures and physiologic limitations, additional constraints were established. First, certain tendon forces were related in the following form:

$$\sum_{j=1}^{n} d_{ij}|\bar{T}_j| = 0 \qquad\qquad i = 1,2,...s \qquad (2)$$

in which

d_{ij} = coefficient for the constraint relationships,

n = number of tendon forces possessing constraint relationships, and

s = number of constraint equations assumed.

Second, the strength of the major active muscles, if further restricted by the physiologic limitations based on the cross-sectional area of each muscle, as analyzed in Chap. 2, can be expressed in the following form:

$$T_j \leq T_j^M \qquad\qquad j = 1, 2, \ldots n \qquad\qquad (3)$$

in which

T_j^M = maximum strength of the muscle limited by its physiological cross-sectional area, and

n = number of constraint conditions for the system.

The constraint equations in the form of equation (2) were obtained based on the relationships of the force distributions among the extensor mechanism. The structures of the dorsal aponeurosis of the human finger are relatively complicated. Its anatomy and function have been subjected to many studies[1,13-17]. However, the exact manner of force transmission between each of the muscles and tendinous bands at the PIP and DIP joints was not available, except for a few special conditions where the joints were at either fully extended or flexed positions. By using strain gauge measurement, Saraffian found a gradual increase in the tension of the extensor slip (ES) beyond 60° of PIP joint flexion, and at 90° flexion of this joint there is total relaxation of the lateral bands[18]. This finding was confirmed by Mick[19]. In spite of the lack of relationship of force distribution among these tendons, the general constraints were established based on the available anatomical knowledge. More specifically, these equations for each finger were assumed as:

Index Finger:

$$
\begin{aligned}
\text{TE} &= \text{RB} + \text{UB} \\
\text{RB} &= 0.667\,\text{LU} + 0.167\,\text{LE} \\
\text{UB} &= 0.333\,\text{UI} + 0.167\,\text{LE} \qquad\qquad (4) \\
\text{ES} &= 0.333\,\text{LU} + 0.167\,\text{LE} + 0.333\,\text{UI} + 0.333\,\text{RI}
\end{aligned}
$$

Long Finger:

$$TE = RB + UB$$
$$RB = 0.133\,RI + 0.167\,LE + 0.667\,LU$$
$$UB = 0.313\,UI + 0.167\,LE \qquad\qquad (5)$$
$$ES = 0.133\,RI + 0.313\,UI + 0.167\,LE + 0.333\,LU$$

Ring Finger:

$$TE = RB + UB$$
$$RB = 0.333\,RI + 0.167\,LE + 0.667\,LU$$
$$UB = 0.200\,UI + 0.167\,LE \qquad\qquad (6)$$
$$ES = 0.333\,RI + 0.200\,UI + 0.167\,LE + 0.333\,LU$$

Little Finger:

$$TE = RB + UB$$
$$RB = 0.317\,RI + 0.167\,LE + 0.667\,LU$$
$$UB = 0.100\,UI + 0.167\,LE \qquad\qquad (7)$$
$$ES = 0.317\,RI + 0.100\,UI + 0.167\,LE + 0.333\,LU$$

Similar constraint conditions were also assumed in the thumb[9] in order to derive the equilibrium equations.

When free-body analysis was carried out for all joints of the finger and the thumb, a total of 18 equilibrium equations were obtained. In addition, four equality constraints in the form of equations (4–7) were included. The total number of unknown variables for the fingers, listed in Table 3-1 and Table 3-2, is 24, which exceeds the number of available equilibrium and constraint equations (a total of 22), thus making the problem indeterminate with 2° of redundancy. Special methods must be adopted to solve this indeterminate problem.

METHODS OF SOLUTION

Two general approaches have been adopted to handle the indeterminate problem derived from the force analysis of the musculoskeletal system. The first method employed was based on the reduction principle (permutation-combination method). With the redundant variables systematically assigned by known forces, the system becomes statically determinate[7,8,9]. Each of the reduced determinate problems was solved uniquely, provided that the coefficient matrix of the system was nonsingular. In addition, those solutions not satisfying the constraint conditions[3] were defined as inadmissible and were subsequently discarded.

An alternate method was introduced based on the principles of optimization[20-23]. This method is briefly reviewed as follows:

Minimize

$$J = f(x_1, x_2, ...x_n) \tag{8}$$

Subject to

$$g_j(x_1, x_2, ...x_n) = 0 , \quad j = 1,2,...m \tag{9}$$

and

$$b_i \leq x_i \leq a_i \qquad i = 1,2,...n \tag{10}$$

where

$$b_i \geq 0$$

When the optimal function, J, and all the constraint equations are linear, linear programming techniques can be applied. Three basic terminologies, which will be beneficial for the following discussion are defined. A *feasible solution* is a set of x_i which satisfies both equality and inequality constraints (equations 9 and 10). A *basic feasible solution* is a feasible solution with no more than m (total number of constraint conditions) non-zero values of x_i (including slack variables, if any). Finally, a basic feasible solution which satisfies the optimal condition[8] and all the constraint equations is defined as an *optimal basic feasible solution*. However, when the nonlinear optimal function, J, is involved, other nonlinear optimization algorithms should be used, and the number of non-zero variables in the solution will no longer be limited by the number of equality equations.

Application of the systematic reduction method serves as a useful asset to the optimization technique, since it produces the basic feasible solution region that encloses all admissible feasible solutions. The advantage of using the optimization technique is that it can produce unique solutions based on different objective criteria, which are based on either physiological expectation or corresponding pathological situations. Unfortunately, all the physiologically related optimal criteria have not been substantiated. Therefore, before any optimal criterion is to be used, a list of all feasible solutions based on the reduction method would be beneficial for general discussion.

In the next section, results for the index finger under various simulated static functions are presented. These functions include basic pinch and grasp, isolated flexion-extension, radial-ulnar deviation, abduction-adduction with the middle finger, and few commonly used daily activities.

These functions were simulated by proper assignment of finger joint orientation at each joint and the externally applied force. The joint orientation angles, which involve only the flexion and extension positions assumed for various functions for all fingers, are summarized in Table 3-3.

Table 3-3 Joint orientation angles for various functions.

Function	Joint Flexion Angle		
	DIP	PIP	MP
Tip pinch	25°	50°	48°
Key pinch	20°	35°	20°
Pulp pinch	0°	50°	48°
Grasp	23°	48°	62°
Abduction	0°	0°	0°
Adduction	0°	0°	0°
Flexion	0°	0°	0°
Extension	0°	0°	45°
Briefcase grip	55°	72°	23°
Holding glass	20°	48°	5°
Opening big jar	35°	55°	50°

The externally applied forces used to simulate these functions are listed in Table 3-4, which includes the magnitude, orientation (in terms of directional cosine), and point of application of these external forces.

In grasp, the externally applied forces at each phalanx were different. The ratio of these forces was determined by the experimental set-up described in Chap. 5. These forces are assumed to apply at the midportion of each phalanx on the volar side and perpendicular to the long axis of the phalanx. Based on experimental tests, six types of phalangeal force distribution are possible in grasp. However, only the results of Type 2 are presented in which the distal phalanx carried the largest force followed by the proximal phalanx and the middle phalanx. The maximum magnitude of the applied force was normalized to be one. Thus, all of the results presented are expressed in terms of the unit of applied force. For obtaining realistic forces and moments in dimensional units, all that is required is multiplying these results with the actual measurement of the externally applied forces, which will be presented in Chap. 5.

Table 3–4 Hypothetically applied forces.

Function	Point of Application (on phalanx)			Orientation of Applied Force (directional cosine)			Magnitude of Applied Force	Phalanx
	X	Y	Z	X	Y	Z		
Tip pinch	−0.55	−0.10	0.0	0.707	0.707	0.0	1.0	distal
Key pinch	−0.20	−0.10	0.0	0.0	0.0	−1.0	1.0	middle
Pulp pinch	−0.55	−0.10	0.0	0.707	0.707	0.0	1.0	distal
Grasp	−0.55	0.0	0.0	0.0	1.0	0.0	1.0	distal
	−0.79	0.0	0.0	0.0	1.0	0.0	0.34	middle
	−1.35	0.0	0.0	0.0	1.0	0.0	0.66	proximal
Abduction	0.0	0.0	0.28	0.0	0.0	−1.0	1.0	middle
Adduction	0.0	0.0	−0.28	0.0	0.0	1.0	1.0	middle
Flexion	−0.44	−0.2	0.0	0.0	1.0	0.0	1.0	distal
Extension	−1.66	0.32	0.0	0.0	−1.0	0.0	1.0	proximal
Briefcase grip	−0.48	0.44	0.0	0.0	1.0	0.0	1.0	middle
Holding glass	−0.44	−0.2	0.0	0.0	1.0	0.0	1.0	distal
Opening big jar	−0.44	−0.2	0.0	0.0	0.707	−0.707	1.0	distal

RESULTS

To facilitate discussion of the results based on linear optimization and systematic reduction methods, the results of tip pinch in the index finger are presented in detail. Similar methods were used in the analysis of the other fingers and the thumb, and their results have been presented elsewhere[7,8,9]. By using the systematic reduction method, 276 sets of feasible solutions which satisfy the equilibrium equation (1) and the equality constraint[4] were obtained. After imposing inequality constraints on muscle and joint forces[3], there are only eight admissible solutions. Each of these eight solutions can be obtained by the linear programming technique upon proper selection of objective functions. Table 3–5 displays the muscle forces corresponding to the eight basic feasible solutions. With these eight known basic feasible solutions, the reduced feasible regions of the redundant system of equations can be defined for two-dimensional space based on any two independent variables.

Table 3-5 Calculated tendon forces of index finger in tip pinch based on permutation-combination method (in units of applied force).

Solution	1	2	3	4	5	6	7	8
TE	0.25	5.88	4.49	4.40	3.96	1.02	0.41	0.34
FP	2.13	6.56	5.46	5.39	5.04	2.73	2.25	2.20
RB	--	2.23	2.47	2.48	2.56	0.84	0.17	0.10
UB	0.25	3.66	2.02	1.92	1.40	0.17	0.24	0.25
FS	1.84	--	--	--	--	1.28	1.73	1.77
ES	0.73	5.39	3.12	2.97	2.25	0.60	0.70	0.71
RI	1.42	5.19	1.42	1.19	--	--	1.14	1.26
LU	--	--	1.85	1.97	2.55	1.26	0.25	0.15
UI	0.77	4.31	2.37	2.25	1.64	0.52	0.72	0.74
LE	--	13.36	7.39	7.02	5.14	--	--	--

These regions are constructed by identifying the optimal basic feasible solutions as unique points in planar space. The adjacent points are connected to form the boundary lines enclosing the feasible region. Such regions can be either bounded or unbounded, depending on the constraint equations. In the case of a linear problem, such as the one we are discussing, the regions will be bounded and also convex.

The reduced feasible solution region can now be plotted based on any two variables in the present case. Two typical reduced feasible solution regions, based on variables sets UI-FP and LU-RB, are shown in Fig. 3–5. They have different shapes and sizes. The points on, or within, the boundary lines

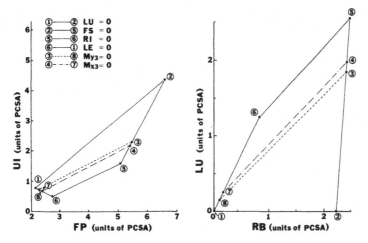

Fig. 3–5 Reduced feasible solution regions based on the variable sets UI-FP and LU-RB.

represent all the feasible solutions. In this case, the eight basic feasible solutions are the intersections of two or more lines of variables with zero values. These eight basic feasible solutions, as identified in Fig. 3–5, are either at the vertices or on the lines. Solution 1 is a degenerative basic feasible solution, since the three independent variables are zero (RB, LU and LE). Basic feasible solutions 3, 4, 7, and 8 are not located at the vertices merely because two constraint moments, Mx_3 and My_3, do not have to satisfy the non-negative conditions, and solutions on either side of the lines, $Mx_3 = 0$ and $My_3 = 0$, will be feasible. These lines are illustrated in Fig. 3–5. However, construction of the feasible solution region does not depend on such variables. Slack variables are introduced to satisfy the non-negative requirement in linear programming formulation.

These eight basic feasible solutions can also be obtained by the optimization method. For example, based on the optimal criteria of minimizing the sum of all tendon forces and the sum of joint constraint moments, the basic feasible solutions 1 and 8 can be obtained, respectively, while certain solutions are unlikely to be obtained by the optimization method unless specific criteria are defined. For example, solutions 3, 4, 7 and 8 are obtained by the four special criteria; namely,

$$J_3 = |My_3| + FS, \quad J_4 = |My_3| + FS, \quad J_7 = |Mx_3| + LE, \quad \text{and} \quad J_8 = |My_3| + LE.$$

These objective functions are difficult to visualize, but they can be identified after the basic feasible solution region is constructed based on the systematic reduction method.

Similarly, multiple solutions are expected for results of other simulated functions. However, only the ranges of these predicted values are summarized in Table 3-6 for muscle forces, Table 3-7 for joint constraint forces, and Table 3-8 for joint constraint moments. The sign convention of the joint forces and moments follows the coordinate system defined in Chap. 1. The results reported here were obtained based on a slightly different constraint condition of muscle relationship, as shown in equation (4).

Table 3-6 Muscle forces of index finger (based on unit applied force).

Function	FP	FS	RI	LU	UI	LE
Tip pinch	1.93~2.08	1.75~ 2.16	0.0~0.99	0.0 ~0.72	0.21~0.65	- -
Pulp pinch	2.53~3.14	0.32~ 1.32	0.0~1.61	0.0 ~1.17	0.62~1.19	- -
Lateral pinch (key)	1.37~5.95	- -	1.01~7.04	0.0 ~6.10	- -	7.45~15.94
Grasp	3.17~3.47	1.51~ 2.14	0.0~1.19	0.0 ~0.91	0.0 ~0.49	- -
Flexion	3.28~4.49	3.57~ 6.43	- -	0.69~1.48	0.20~1.06	0.0 ~ 5.34
Extension	0.0 ~1.08	- -	0.0~0.45	0.0 ~0.15	0.77~1.17	12.12~14.84
Abduction	0.0 ~1.20	0.0 ~ 3.76	5.0~5.26	0.0 ~0.02	- -	3.87~ 7.24
Adduction	0.0 ~6.57	0.0 ~22.4	0.0~2.37	- -	5.0 ~7.60	10.45~33.6
Briefcase grip	0.0 ~0.02	1.70~ 1.78	0.0~0.45	0.0 ~0.33	0.11~0.27	- -
Holding glass	2.77~2.99	1.29~ 1.57	- -	0.48~0.53	0.28~0.38	- -
Opening jar	3.50~5.49	- -	4.2~4.53	0.0 ~1.15	0.0 ~1.0	9.48~16.23

DISCUSSION

Although the issue of selecting optimal criteria in solving the statically indeterminate problems in biomechanics is unsettled, the present analysis can help in the construction of certain objective concepts to stimulate further investigation. The multiplicity of feasible solutions reflects the fact that there would be no unique solution for the redundant problem whenever the applied strength to the anatomical system is below its physiological limit. Under such a circumstance, all muscles can contribute at a varying magnitude to minimize fatigue, and at the same time, the total energy output could also be conserved. A number of preferable muscle force distribution patterns may exist, but a unique set of fixed muscular contributions would be unlikely to occur since the muscular system has many different options for providing required strength, particularly when the anatomical structure is under isometric resistance for a prolonged period of time. This phenomenon has been observed from electromyographic (EMG) evaluations of skeletal muscles under normal loads, in

Table 3-7 Joint constraint forces of index finger.

Function	CX2	CX4	CX6	CY2	CY4	CY6	CZ2	CZ4	CZ6
Tip pinch	-2.4~ 2.7	-4.4~ 4.9	-3.5~ 3.9	0.2~ 0.3	0.9~ 1.1	2.1~ 2.3	-0.1~ -0.1	-0.0~ -0.0	0.1~ 0.2
Key pinch	-2.9~ 12.5	-4.9~ -19.4	-14.7~ -27.1	0.7~ 3.2	1.1~ 4.5	3.9~ 5.7	0.7~ 0.9	0.3~ 1.1	0.0~ 0.1
Pulp pinch	-3.0~ 4.6	-4.8~ 5.8	-4.0~ 4.6	-0.0~ -0.2	1.1~ 1.4	2.2~ 2.4	-0.1~ -0.2	0.0~ 0.0	0.1~ 0.1
Grasp	-2.8~ 3.4	-4.5~ 5.3	-3.2~ 3.7	0.5~ 0.7	1.0~ 1.3	2.9~ 3.1	-0.2~ -0.2	-0.0~ -0.1	0.3~ 0.4
Abduction	0.0~ 2.4	-3.0~ 7.4	-9.5~ -15.7	0.0~ 0.3	0.2~ 0.4	1.7~ 1.8	0.0~ -0.1	0.9~ 1.1	0.4~ 0.4
Adduction	0.0~ 12.9	18.2~ 48.6	23.0~ 66.8	0.0~ 1.7	1.2~ 2.7	1.4~ 2.9	0.0~ -0.3	-0.4~ -0.8	-0.8~ -1.1
Flexion	3.2~ 5.6	8.1~ 16.1	-7.9~ -18.4	0.0~ -0.3	0.0~ -0.4	0.5~ 1.1	-0.2~ -0.2	0.0~ -0.0	0.1~ 0.2
Extension-3rd phalanx	6.9~ 8.3	0.0~ 9.7	15.0~ 16.8	0.9~ 1.1	0.6~ 0.7	4.4~ 4.9	-0.2~ -0.2	-0.1~ -0.3	0.2~ 0.2
Briefcase grip	0.0~ 0.0	1.7~ 1.9	1.0~ 1.3	0.0~ 0.0	0.0~ 0.2	0.6~ 0.7	0.0~ -0.0	0.0~ 0.0	0.0~ 0.0
Holding drinking glass	2.5~ 2.9	4.3~ 4.4	4.0~ 4.1	0.2~ 0.3	1.1~ 1.1	0.9~ 0.9	-0.2~ -0.2	-0.0~ -0.1	0.2~ 0.2
Opening big jar	5.2~ 9.5	7.2~ 14.2	14.8~ 24.3	1.7~ 3.3	2.4~ 4.9	6.5~ 9.9	0.3~ 0.5	0.2~ 0.8	0.2~ 0.3

Table 3-8 Joint constraint moments of index finger.

Function	MX2	MX4	MX6	MY2	MY4
Tip pinch	-0.007	-0.01 ~ -0.006	-0.28 ~ -0.3	-0.074~ -0.076	0.0 ~ -0.05
Key pinch	-0.1 ~ -0.1	-0.1 ~ -0.6	-3.0 ~ -3.3	0.3 ~ 0.4	0.0 ~ 2.2
Pulp pinch	0.006~ 0.01	-0.004~ 0.07	-0.26 ~ -0.29	-0.09 ~ -0.1	-0.05~ 0.2
Grasp	-0.01	-0.04 ~ 0.01	-0.5 ~ -0.6	-0.12 ~ -0.13	0.0 ~ -0.2
Abduction	0.0 ~ 0.005	-0.03 ~ 0.04	-1.0 ~ -1.1	0.0 ~ -0.02	0.0 ~ 0.6
Adduction	0.0 ~ 0.03	0.1 ~ 0.3	0.0 ~ 0.6	0.0 ~ -0.1	0.0 ~ 1.9
Flexion	0.008~ 0.014	0.01 ~ 0.03	0.0 ~ -0.4	-0.1 ~ -0.2	-0.09~ 0.016
Extension-3rd phalanx	0.016~ 0.019	0.02 ~ 0.13	-0.2 ~ -0.3	-0.07 ~ -0.08	0.0 ~ 1.0
Briefcase grip	0.0	0.0 ~ -0.005	-0.07 ~ -0.08	0.0	0.0 ~ 0.01
Holding glass	-0.006~ -0.007	0.010~ -0.011	-0.152~ -0.153	-0.107~ -0.110	0.0 ~ -0.07
Opening big jar	-0.15 ~ 0.15	-0.1 ~ -0.8	-3.0 ~ -3.2	0.36 ~ 0.38	0.0 ~ 1.9

which the EMG signals of certain muscles were found to be highly variable. Under such circumstances, all basic feasible solutions, and even some of the feasible solutions, are possible. This hypothesis helps to explain why marginal reductions of muscle strength due to neurological deficits would generally produce small effects on the less powerful functions of the joint system. However, certain joint deformities are commonly observed due to muscle imbalance in the neutral state.

In general, the variation of tendon and joint forces is only marginal based on all optimal criteria used, particularly for joint constraint forces. Therefore, the mean values of tendon and joint forces can serve as the representative values of finger internal forces in a normative model. Especially when these values are expressed on a relative basis, the ratio will become more consistent, regardless of the form of optimal criteria necessary for the solution.

The flexor profundus and sublimis have consistently high force values as compared to other muscles in most of the functions. In lateral key pinch, the flexor sublimis carries no load, however, the long extensors and two intrinsic muscles on the radial side contribute large forces. In key pinch, the balance of the MP joint's contribution in ulnar deviation is mainly attributed to the action of the radial interosseous, which also creates a large flexion moment at the MP joint. The counteraction of the long extensor in balancing this moment is, therefore, expected. The intrinsic muscles usually produce more forces during pinch functions than those of grasp due to the stabilization requirement of the MP joint.

During the four isolated finger functions, namely, flexion, extension, abduction and adduction, those muscles having the most favourable anatomical position for these functions generate the most forces. The muscle forces in simulated daily activities are quite similar to those of the basic pinch functions.

Joint constraint forces of pinch functions are also of interest to discuss. The joint contact forces (F_x) are comparable for both the PIP and MP joints. In lateral pinch, such force becomes significantly larger in the MP joint as compared to the PIP joint. Since pinch forces are substantially large for tip and pulp pinches, the normal contact force at the PIP joint cannot be overlooked. Almost all of the volar-dorsal shear forces are in the dorsal direction with substantial magnitudes. This implies that under such activities all fingers have a tendency toward volar subluxation (the distal bone tends to move downward against the bone). The radial-ulnar shear forces at the MP joint are almost in the radial direction. This suggests that the proximal phalanx has a tendency to load the support structures on the radial side and also shift in the ulnar direction. These results have significant clinical implications in treating joint instability and in understanding joint deformities. If prosthetic joints are being used, adequate constraint has to be introduced if the soft tissues are unable to maintain the constraining function.

The results shown in Table 3–8 reveal moderate amounts of internal or external axial rotation moments at all three joints when a finger is performing pinch actions. Radial or ulnar bending moments also exist at two interphalangeal joints. Such results demonstrate the need of a truly three-dimensional analysis. With two-dimensional analysis, such out-of-plane forces and moments can never be realized. In the study of finger joint pathomechanics and the design of joint prosthetic implants, such data are of paramount importance.

The results being discussed thus far were obtained based on the mathematical model and are expressed in dimensionless form. With the additional information obtained from functional strength measurements, internal joint and tendon forces can now be converted into realistic forces in metric units (Tables 3–9 and 3–10). Such dimensional values can help to provide the strength requirements for prosthetic joint and tendon design. Bench test criteria can also be established to accomplish the necessary laboratory evaluation of these artificial implant devices.

The only available direct measurements of *in vivo* tendon forces under isometric function were studied by Bright and Urbaniak[24] in a very limited number of subjects under local anesthesia during carpal tunnel procedures. In their preliminary report, flexor tendon tension was found to be in the range of 2.5–12.5kg for the FP and 1.0–7.5kg for the FS in tip pinch. In grasp, tendon forces fell into a range of 4.0–20.0kg and 1.25–15.0kg for the FP and FS, respectively. These values are relatively lower than our theoretically predicted results. Such discrepancies are expected since the denervation of the median nerve during carpal tunnel procedures would partially inactivate the thenar muscles which will weaken pinch and grasp functions significantly. Reliable verification of the theoretical results presented here awaits the development of better experimental methods. A qualitative comparison can be achieved through quantitative measurements of muscle EMGs, which will be discussed in Chap. 7.

This chapter presents a generalized theory on muscle force distributions under selected isometric functions. The methods to resolve the formulated indeterminate problems are systematic reduction (permutation-combination method) and linear and nonlinear optimization methods. The results indicate that the joint configurations, the anatomical structures of the muscles, and the manner of external load application are the primary factors of the distribution of the internal forces. Although the results presented are based on the index finger only, similar results are available for the other fingers and the thumb by using the normative model described in Chap. 1. The discussion and inter-

Table 3-9   Muscle forces of index finger (kg).

Function	FP	FS	RI	LU	UI	LE
Tip pinch	10.04 ~ 10.82	9.10 ~ 11.23	0.0 ~ 5.15	0.0 ~ 3.74	1.09 ~ 3.38	– –
Pulp pinch	12.90 ~ 16.01	1.63 ~ 6.73	0.0 ~ 8.21	0.0 ~ 5.97	3.16 ~ 6.07	– –
Grasp	22.19 ~ 24.29	10.57 ~ 14.98	0.0 ~ 8.33	0.0 ~ 6.37	0.0 ~ 3.43	– –
Flexion	18.24 ~ 24.96	19.85 ~ 35.75	– –	3.84 ~ 8.23	1.11 ~ 5.89	0.0 ~ 29.69
Extension	0.0 ~ 2.95	– –	0.0 ~ 1.23	0.0 ~ 0.41	2.10 ~ 3.19	33.09 ~ 40.51
Abduction	0.0 ~ 4.08	0.0 ~ 12.78	17.0 ~ 17.88	0.0 ~ 0.07	– –	13.16 ~ 24.62
Adduction	0.0 ~ 19.05	0.0 ~ 64.96	0.0 ~ 6.87	– –	14.5 ~ 22.04	30.31 ~ 97.44

Table 3-10   Joint constraint forces of index finger (kg).

Function	-CX2	-CX4	-CX6	CY2	CY4	CY6	CZ2	CZ4	CZ6
Tip pinch	12.4 ~ 14.1	22.7 ~ 25.5	18.1 ~ 20.3	1.2 ~ 1.7	4.8 ~ 5.7	10.8 ~ 12.0	-0.6 ~ -0.7	-0.1 ~ -0.2	0.8 ~ 0.9
Pulp pinch	15.3 ~ 23.2	24.6 ~ 29.6	20.6 ~ 23.6	-0.1 ~ -1.1	5.6 ~ 7.1	11.1 ~ 12.5	-0.6 ~ -0.8	0.0 ~ -0.2	0.4 ~ 0.6
Grasp	19.5 ~ 24.1	31.7 ~ 36.8	22.7 ~ 26.2	3.4 ~ 4.6	7.3 ~ 8.8	20.5 ~ 22.0	-1.4 ~ -1.5	-0.3 ~ -0.6	2.2 ~ 2.6
Flexion	17.8 ~ 31.0	45.0 ~ 89.5	44.1 ~ 102.3	0.0 ~ -1.8	0.0 ~ -2.5	2.8 ~ 6.0	-1.0 ~ -1.3	0.0 ~ -0.1	0.8 ~ 1.3
Extension	18.9 ~ 22.5	0.0 ~ 26.5	41.1 ~ 46.0	2.5 ~ 3.0	1.7 ~ 1.8	11.9 ~ 13.3	-0.5 ~ -0.6	-0.3 ~ 0.7	0.5 ~ 0.6
Abduction	0.0 ~ 8.0	10.3 ~ 25.0	32.3 ~ 53.4	0.0 ~ 1.1	0.7 ~ 1.4	5.9 ~ 6.0	0.0 ~ -0.2	3.0 ~ 3.6	1.3 ~ 1.4
Adduction	0.0 ~ 37.3	52.8 ~ 141.0	66.7 ~ 193.8	0.0 ~ 5.0	3.5 ~ 7.9	3.9 ~ 8.5	0.0 ~ -0.9	-1.3 ~ -2.4	-2.3 ~ -3.3

pretation of the data will be more apparent when the simulations of the abnormal conditions are discussed.

REFERENCES

1. Smith, E.M., Juvinall, R.C., Bender, L.F. and Pearson, J.R., "Role of the finger flexors in rheumatoid deformities of the metacarpophalangeal joints", *Arthritis Rheum.*, **7**:467–480 (1964).

2. Thomas, D.H., Long, C. and Landsmeer, J.M.F., "Biomechanical considerations of lumbricalis behavior in the human finger", *J. Biomech.*, **1**: 107–115 (1968).

3. Hirsch, D., Page, D., Miller, D., Dumbleton, J.H. and Miller, E.H., "A biomechanical analysis of the metacarpophalangeal joint of the thumb", *J. Biomech.*, **7**: 343–348 (1974).

4. Storace, A. and Wolf, B., "Functional analysis of the role of the finger tendons", *J. Biomech.*, **12**: 575–578 (1979)

5. Berme, N., Paul, J.P. and Purvis, W.K., "A biomechanical analysis of the metacarpophalangeal joint", *J. Biomech.*, **10**: 409–412 (1977).

6. Toft, R. and Berme, N., "A biomechanical analysis of the joints of the thumb", *J. Biomech.*, **13**: 353–360 (1980).

7. Chao, E.Y., Opgrande, J.D. and Axmear, F.E., "Three-dimensional force analysis of finger joints in selected isometric hand functions", *J. Biomech.*, **9**: 387–396 (1976).

8. Cooney, W.P. III and Chao, E.Y., "Biomechanical analysis of static forces in the thumb during hand functions", *J. Bone Joint Surg.*, **59-A**: 27–36 (1977).

9. Chao, E.Y. and Cooney, W.P. III "Internal Forces in Normal Hand", Chap. 2, pp. 99–109, in *Human Joints and Their Artificial Replacements*, P. Walker, ed. (Chas. C. Thomas, Springfield, IL, First edn. 1977).

10. Chao, E.Y. and An, K.N., "Determination of internal forces in human hand", *J. Eng. Mech., Div. ASCE.*, **104**: 255–272 (1978).

11. Chao, E.Y. and An, K.N., "Graphical interpretation of the solution to the redundant problem in biomechanics", *J. Biomech. Eng.*, **100**: 159–167 (1978).

12. An, K.N., Chao, E.Y., Cooney, W.P. III and Linscheid, R.L., "Normative model of human hand for biomechanical analysis", *J. Biomech.*, **12**: 775–788 (1979).

13. Tubiana, R. and Valentin, P., "The physiology of the extension of the fingers", *Surg. Clin. No. Am.*, **44**: 907–918 (1964).

14. Landsmeer, J.M.F., "The anatomy of the dorsal aponeurosis of the

human finger and its functional significance", *Anat. Rec.,* **104**: 31 (1949).

15. Stack, H.G., "Muscle function in the finger", *J. Bone Joint Surg.,* **44-B**: 899 (1962).

16. Harris, C. and Rutledge, A.L., "The functional anatomy of the extensor mechanism of finger", *J. Bone Joint Surg.* **54-A**: 713-726 (1972).

17. Haines, R.W., "The extensor apparatus of the finger", *J. Anat.* **85**: 251 (1951).

18. Saraffian, S.K., Kazarian, L.E. and Topouzian, L.K. *et al.,* "Strain variation in the components of the extensor apparatus of the finger during flexion and extension. A biomechanical study", *J. Bone Joint Surg.,* **52-A**: 980-990 (1970).

19. Mick, J., "Confirmation of differential loading of lateral and central fibers of the extensor tendon", *J. Hand Surg.,* **6**: 462-467 (1981).

20. Barbenel, J.C., "Biomechanics of the temporomandibular joint. A theoretical study", *J. Biomech.,* **5**: 251-256 (1972).

21. Nubar, Y. and Contini, R., "A minimal principal in biomechanics," *Bull. Math. Biophys.,* **23**: 377 (1961).

22. Seireg, A. and Arvikar, R.J., "A mathematical model for evaluation of forces in lower extremities of musculoskeletal systems", *J. Biomech.,* **6**: 313-326 (1973).

23. Penrod, D.D., Davy, D.T. and Singh, D.P., "An optimization approach to tendon force analysis". *J. Biomech.,* **7**: 123-129 (1974).

24. Bright, D.S. and Urbaniak, J.B., "Direct measurements of flexor tendon tension during active and passive digit motion and its application to flexor tendon surgery", *Trans. of 22nd Ann. Mtg., Orthopaedic Research Society,* page 240, (1976).

4

HAND JOINT ORIENTATION AND RANGE OF MOTION

The practical application of hand joint orientation and range of motion measurements encompasses a major aspect of assessing the functional improvement in arthritic patients. Effective joint motion measuring techniques combined with hand strength and dexterity tests will provide the objective means for physicians to evaluate the outcome of their treatments. This information can also be applied to prosthetic joint design and testing. Since finger and thumb joints are small and their motion complex, the evaluation technique must be accurate, reproducible and easy-to-use. These requirements are met by the method to be described in this chapter.

INTRODUCTION

After the hip and knee joints, degenerative and rheumatoid arthritis are most commonly observed in the hand[1,2,3]. The significance of these problems is particularly apparent in the fingers and thumb, where arthritic deformity can cause significant loss of the prehensile hand functions of both pinch and grasp[4,5]. Effective use of the hands becomes difficult as structural instability, pain and loss of motion progresses.

There is little quantitative information upon which the functional range of motion, strength and dexterity of diseased or injured fingers and thumb can be compared. In the thumb, motion in the planes of flexion-extension and abduction-adduction can be recorded at the interphalangeal and metacarpophalangeal joints as long as these motions are occurring individually from their neutral position. If such motions take place simultaneously or the intricate movement occurs at a more complex joint, such as thumb basal joint, it is difficult to quantitate their precise kinematic behavior three-dimensionally. Such information could be estimated based on thumb position[6,7]. Pieron[12] performed a classic and detailed study of the thumb carpometacarpal joint based on biplanar vector analysis for cadaver specimens. A thumb goniometer was proposed, but its clinical applicability remains to be established. In normal fingers, distal and proximal interphalangeal joints have exclusive flexion-extension motion which can be easily measured. However, the MCP joint becomes more complex with multiple axes of rotation. Its true motion can only be approximated, particularly when such compound motion takes place simultaneously. When an abnormal hand with pronounced joint deformity is involved, such measurement becomes even more difficult.

Loss of motion, instability and pain are recognizable clinical findings in major forms of arthritis. However, to quantitate the joint deformities and loss of motion caused by pathological changes was not possible in the past. Therefore, such symptoms could only be assessed subjectively. Furthermore, such deformities and loss of range of motion have never been correlated with the dexterity and functional strength of the hand. The lack of such knowledge has obvious effects on the development of better methods of treatment and rehabilitation for diseased and disabled hand patients.

The purpose of this chapter is to report the course of development of an objective method to measure the three-dimensional joint orientation and range of motion in fingers and thumb and to use such measurements as objective criteria to assess finger and thumb function impairments. Successful application of this method can achieve three specific aims:

a) to quantitate joint angulation in the fingers and thumb under common functional configurations,

b) to measure the joint range of motion as obtained from the extreme positions of the fingers and thumb, and

c) to assess the dexterity and coordination of the fingers and thumb in timed peg-board manipulation.

The information about joint orientation of hand functions was later used to calculate the internal joint and tendon forces.

EXPERIMENTAL METHOD

The range of joint motion has been traditionally measured by a hand-held goniometer technique. Such a method is not only imprecise but also unable to determine rotation or abduction-adduction movement, particularly for badly deformed hands. A new method was developed to quantitate the three-dimensional motion in hand joints. This technique utilizes biplanar radiographic recording of the spatial position and orientation of special markers fixed to each mobile finger and thumb segment (Fig. 4–1). The markers are "T" shaped, and when they are properly placed on the dorsum of the finger and thumb segments, reference coordinate axes can be defined. When the orientation of the coordinate axes for each bony element of the hand is defined with respect to a common inertial reference frame, the relative joint angles can

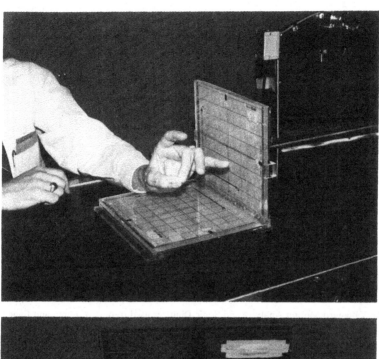

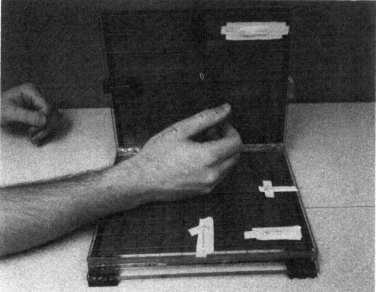

Fig. 4-1   Biplanar X-ray method used to determine the accurate three-dimensional finger and thumb joint orientation in various hand functional configurations.

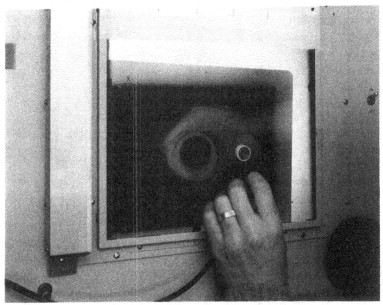

Fig. 4–2   Vanguard Motion Analyzer and Sonic Digitizer used to obtain thumb and finger joint angles and range of motion.

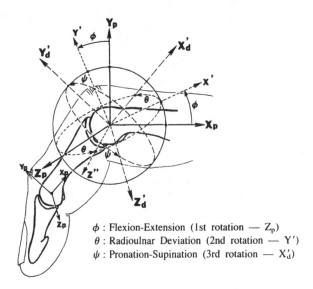

ϕ : Flexion-Extension (1st rotation — Z_p)
θ : Radioulnar Deviation (2nd rotation — Y')
ψ : Pronation-Supination (3rd rotation — X'_d)

Fig. 4–3   Joint orientation and rotational sequence of a finger or thumb joint.

be determined uniquely. Thumb and finger (index, long, ring and little) joint orientations are then determined for the functional hand configurations of pinch, grasp and extremes of flexion-extension, abduction-adduction, and opposition. The X-ray records of each position are measured by a Vanguard Motion Analyzer with the aid of a Sonic Digitizer (Fig. 4-2). An analytical method derived for the study of elbow motion was adopted[8]. The orientation of finger joints and the distal thumb joints are defined according to Fig. 4-3. Since the thumb basal joint involves a small bone (trapezium) with different anatomical structure (saddle geometry) and complex reference planes, new joint motion definitions are used for the thumb base (Fig. 4-4). Accurate measurements of angular orientation based on the present method has been established to within an error margin of ±4 degrees[8].

The assessment of hand functional dexterity was accomplished by using timed peg-board test (Fig. 4-5). Four types of pegs (cylinder, block, screw and key lock) were transferred from one row to another with the required time recorded by a stopwatch. The evaluation provides an indirect assessment of the manipulative ability of the hand. The time required for such manipulation will be scaled later to form an assessment score.

Trapeziometacarpal Axis System

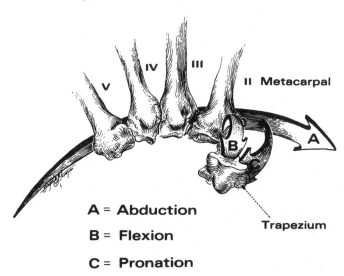

A = Abduction
B = Flexion
C = Pronation

Fig. 4-4 Bony orientation of trapezium and the motion of the CMC joint with respect to the metacarpal bones of the palm of the hand.

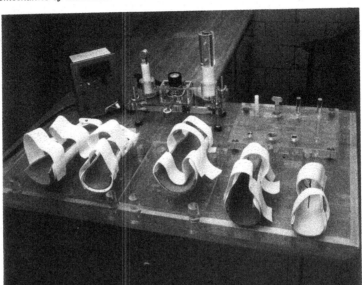

Fig. 4-5 Hand strength evaluation instruments and the timed peg-board functional test for the dexterity of the thumb and opposing fingers.

A composite computerized evaluation form to incorporate joint range of motion, manipulative ability and functional strength along with clinical findings of synovitis, tenderness, joint instability and pain in abnormal patients will be used to compile the test data. Correlation of clinical and radiographic findings with joint functional impairment is now possible. Such information will help to evaluate the result of treatment and rehabilitation.

ANALYTICAL METHODS

A.   Spatial Location of the Marker

The spatial location of any point, p, can be determined based on the images of the marker projected on the two orthogonal cassettes. If the focal points of the X-ray beams have locations $P(X'_1, Y'_1, Z'_1)$ and $Q(Y'_2, Y'_2, Z'_2)$ based on the grid system (X', Y', Z'), the location of point p can be obtained from the intersection of the two beams $(\overline{PS}_1$ and $\overline{QS}_2)$ through the marker image (Fig. 4-6). Due to anticipated experimental error, the two rays PS_1 and QS_2 may not intersect. Then, the midpoint of the unique common perpendicular line, S, between $\overline{PS}_1$ and $\overline{QS}_2$ can be used to approximate the position of point p. The position vector of p in the grid system (X', Y', Z') can be derived as follows:

$$\bar{r}_p = \bar{r}_{Q/o} + \bar{r}_{T_2/Q} - \frac{S}{2}\,\bar{n} \tag{1}$$

where S = length of the common perpendicular line,

$\bar{n}$ = unit vector along the common perpendicular, and

$\bar{r}$ = position vector of any point with respect to the origin,

and

$$\bar{r}_{T_2/Q} = L_2\bar{e}_2 \tag{2}$$

where

$$L_2 = \frac{(\bar{r}_{Q/P} \cdot \bar{e}_1)(\bar{e}_1 \cdot \bar{e}_2) - (\bar{r}_{Q/P} \cdot \bar{e}_2)}{1 - (\bar{e}_1 \cdot \bar{e}_2)^2} \tag{3}$$

The unit vectors $\bar{e}_1$ and $\bar{e}_2$ are defined in Fig. 4–6.

This technique provides accurate spatial locations of the markers over the bones of the thumb and fingers. This information helps to establish the cartesian coordinate axes fixed to each long bone of the hand from which the relative orientation angles of the digital joints can be determined.

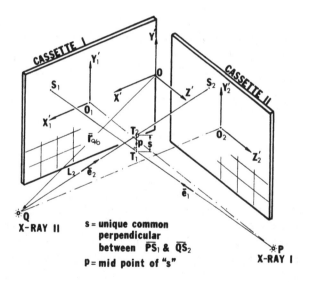

Fig. 4-6 Determination of unique position of marker point in the three-dimensional space defined by the orthogonal cassette using biplanar X-ray technique.

B. *Determination of the Eulerian Angles*

Eulerian angles are often used to describe the orientation of a rigid body in space. Since finite angular displacement does not obey the laws of vector addition, the order in which rotation occurs is important. The rotational sequence which corresponds to the anatomical motion of the finger and thumb joints is shown in Fig. 4–3. The first rotation, ϕ, occurs about the z-axis of the local system which represents flexion-extension. The second rotation, θ, about the x-axis represents abduction-adduction, and the last rotation, ψ, represents internal and external rotation, or pronation-supination. For the CMC joint of the thumb base, the respective axes of rotation are shown in Fig. 4–7. The two coordinate systems at either side of the joint can be transformed following the matrix relationship:

$$\begin{bmatrix} \bar{i} \\ \bar{j} \\ \bar{k} \end{bmatrix} \begin{bmatrix} c\phi c\theta & s\phi c\theta & -s\theta \\ -s\phi c\psi + c\phi s\theta s\psi & c\phi c\psi + s\phi s\theta s\psi & c\theta s\psi \\ s\phi s\psi - c\phi s\theta c\psi & -c\phi s\psi + s\phi s\theta c\psi & c\theta c\psi \end{bmatrix} \begin{bmatrix} \bar{I} \\ \bar{J} \\ \bar{K} \end{bmatrix} \tag{4}$$

where

$\qquad$ c = cosine, and

$\qquad$ s = sine.

$\bar{i}, \bar{j}, \bar{k}$ are the unit base vectors along the distal axes of the joint and $\bar{I}, \bar{J}, \bar{K}$ coincide with the axes fixed to the proximal bony element. If $\bar{i}, \bar{j}, \bar{k}$ and $\bar{I}, \bar{J}, \bar{K}$ are defined in terms of the global system (the grid system, X', Y', Z'), then the individual Eulerian angles can be determined from the following simple equations.

$$\theta = \sin^{-1}(\bar{i} \cdot \bar{K}) \tag{5}$$

$$\phi = \sin^{-1} \frac{\bar{i} \cdot \bar{J}}{\cos\theta} \tag{6}$$

$$\psi = \sin^{-1} \frac{\bar{j} \cdot \bar{K}}{\cos\theta} \tag{7}$$

For every joint position, these angles are unique. Thus, they provide a consistent definition in describing joint position. When consecutive measurements are obtained between extreme limits of joint motion, its range of movement can also be determined. Although this is a rather awkward means to define joint motion in contrast to common clinical definitions, few other options exist to achieve the purpose since three-dimensional motion must follow the principles of rigid body kinematics.

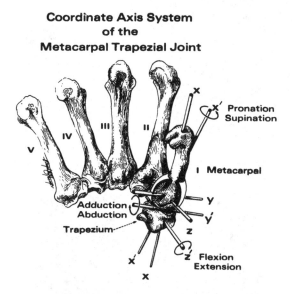

Fig. 4-7 Joint orientation and axes of rotation of the thumb basal joint.

C. Determination of Thumb Carpometacarpal (CMC) Joint Orientation

The motion definition and axes of reference for the CMC joint of the thumb have been described in Fig. 4-7. Because of the anatomical limitation existing in the CMC joint, the above described method cannot be applied directly. The carpal bone is irregular and too small to allow superficial attachment of "T" markers. Thus, a modified method has to be used in which the fixed relationship between the third metacarpal and the trapezium in normal cadaver specimens was studied and quantitated first. Since the third metacarpal and trapezium have very little relative motion, such angulation should remain fixed throughout the range of CMC joint motion. Therefore, if "T" markers are used on both the first and the third metacarpals, the exact CMC joint orientation can be defined by an indirect transformation based on the established fixed orientation between the third metacarpal and the trapezium.

Ten marked cadaver specimens were used to specify the angulation of the trapezium relative to the third metacarpal. Specially constructed "T" markers were drilled into the body of the trapezium and fixed firmly onto the dorsum of the third metacarpal (Figs. 4-8 and 4-9). The exact orientation between these two bones could then be determined from marker points projected on biplanar roentgenograms following the same analytical method described

 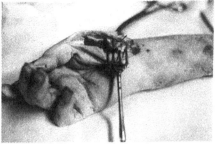

Fig. 4–8a "T" markers fixed on the dorsum of the 2nd and 3rd metacarpal bones.

Fig. 4–8b "T" marker drilled into the trapezium.

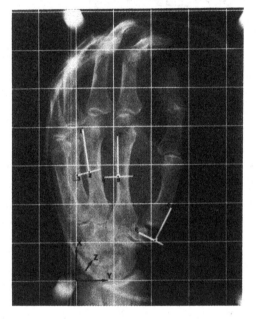

Fig. 4–9 X-ray view of the "T" markers in position for thumb basal joint orientation study.

above. The reference system used for the biplanar X-ray analysis is illustrated in Fig. 4–10.

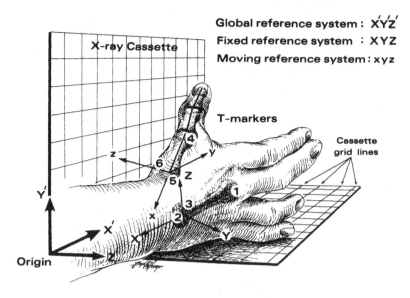

Fig. 4-10 Reference axes as defined by the "T" markers for the thumb segments and the third metacarpal.

The following analytical derivation is used to relate the third metacarpal to the first metacarpal in order to determine the CMC joint orientation. Assume that

$\{\bar{i}_2, \bar{j}_2, \bar{k}_2\}$ = unit vector fixed to the first metacarpal,

$\{\bar{i}_1, \bar{j}_1, \bar{k}_1\}$ = unit vector fixed to the trapezium, and

$\{\bar{I}, \bar{J}, \bar{K}\}$ = unit vector fixed to the third metacarpal.

Then, the following transformation relationship exists.

$$
\begin{bmatrix} \bar{i}_2 \\ \bar{j}_2 \\ \bar{k}_2 \end{bmatrix} = \begin{bmatrix} c\phi_2 c\theta_2 & s\phi_2 c\theta_2 & -s\theta_2 \\ -s\phi_2 c\psi_2 + c\phi_2 s\theta_2 s\psi_2 & c\phi_2 c\psi_2 + s\phi_2 s\theta_2 s\psi_2 & c\theta_2 s\psi_2 \\ s\phi_2 s\psi_2 + c\phi_2 s\theta_2 c\psi_2 & -c\phi_2 s\psi_2 + s\phi_2 s\theta_2 c\psi_2 & c\theta_2 c\psi_2 \end{bmatrix} \begin{bmatrix} \bar{I} \\ \bar{J} \\ \bar{K} \end{bmatrix}
$$

$$
= [\Phi_2] \begin{bmatrix} \bar{I} \\ \bar{J} \\ \bar{K} \end{bmatrix}
\tag{8}
$$

ϕ_2, θ_2 and ψ_2 are the relative angles between the first metacarpal and the third metacarpal which can be determined by the biplanar X-ray method. If one knows the fixed angular orientation between the trapezium and the third metacarpal (which was obtained experimentally from cadaver specimens) expressed by angles ϕ_1, θ_1 and ψ_1, then the following transformation also holds.

$$\begin{bmatrix} \bar{i}_1 \\ \bar{j}_1 \\ \bar{k}_1 \end{bmatrix} = [\Phi_1] \begin{bmatrix} \bar{I} \\ \bar{J} \\ \bar{K} \end{bmatrix} \tag{9}$$

where $[\Phi_1]$ has the same form as Φ_2 except that ϕ_2, θ_2 and ψ_2 angles are replaced by ϕ_1, θ_1 and ψ_1 respectively. These new angles (ϕ_1, θ_1 and ψ_1) describe the rotational transformation from the trapezium system to the third metacarpal system. From equation (9)

$$\begin{bmatrix} \bar{I} \\ \bar{J} \\ \bar{K} \end{bmatrix} = [\Phi_1]^{-1} \begin{bmatrix} \bar{i}_1 \\ \bar{j}_1 \\ \bar{k}_1 \end{bmatrix} \tag{10}$$

Substitute equation (10) into equation (8)

$$\begin{bmatrix} \bar{i}_2 \\ \bar{j}_2 \\ \bar{k}_2 \end{bmatrix} = [\Phi_2][\Phi_1]^{-1} \begin{bmatrix} \bar{i}_1 \\ \bar{j}_1 \\ \bar{k}_1 \end{bmatrix} \tag{11}$$

But

$$\begin{bmatrix} \bar{i}_2 \\ \bar{j}_2 \\ \bar{k}_2 \end{bmatrix} = [\Phi] \begin{bmatrix} \bar{i}_1 \\ \bar{j}_1 \\ \bar{k}_1 \end{bmatrix} \tag{12}$$

where

$[\Phi] = [\Phi_2]$ with ϕ_2, θ_2, ψ_2 replaced by ϕ, θ, ψ,

ϕ = relative flexion-extension of first metacarpal w.r.t. the
   trapezium

θ = relative radio-ulnar deviation angle of first metacarpal w.r.t. the trapezium, and

ψ = relative axial rotation (pronation-supination) of the first metacarpal w.r.t. the trapezium.

The objective now is to determine the orientation angles ϕ, θ and ψ with ϕ_1, θ_1, ψ_1, ϕ_2, θ_2, ψ_2 all known, which can be achieved based on equation (11). Following the similar expression shown in equation (12), one can write

$$\theta = -\sin^{-1}(\bar{i}_2 \cdot \bar{k}_1) \tag{13}$$

$$\psi = \sin^{-1}\frac{\bar{j}_2 \cdot \bar{k}_1}{\cos\theta}, \text{ or} \tag{14}$$

$$\psi = \cos^{-1}\left(\frac{\bar{k}_2 \cdot \bar{k}_1}{\cos\theta}\right) \tag{15}$$

$$\phi = \sin^{-1}\frac{\bar{i}_2 \cdot \bar{j}_2}{\cos\theta}, \text{ or} \tag{16}$$

$$\phi = \cos^{-1}\frac{\bar{i}_2 \cdot \bar{j}_1}{\cos\theta}, \text{ or} \tag{17}$$

From equation (11), since $[\Phi_1]$ and $[\Phi_2]$ are orthogonal matrices, thus

$$[\Phi_1]^{-1} = [\Phi_1]^T = [\Psi_1]$$

where

$$[\Psi_1] = \begin{bmatrix} c\phi_1 c\theta_1 & -s\phi_1 c\psi_1 + c\phi_1 s\theta_1 s\psi_1 & s\phi_1 s\psi_1 + c\phi_1 s\theta_1 c\psi_1 \\ s\phi_1 c\theta_1 & c\phi_1 c\psi_1 + s\phi_1 s\theta_1 s\psi_1 & -c\phi_1 s\psi_1 + s\phi_1 s\theta_1 c\psi_1 \\ -s\theta_1 & c\theta_1 s\psi_1 & c\theta_1 c\psi_1 \end{bmatrix} \tag{18}$$

Therefore

$$\begin{bmatrix} \bar{i}_2 \\ \bar{j}_2 \\ \bar{k}_2 \end{bmatrix} = [\Psi_2][\Psi_1]\begin{bmatrix} \bar{i}_1 \\ \bar{j}_1 \\ \bar{k}_1 \end{bmatrix} \tag{19}$$

Let

$$[A] = [\Phi_2][\Psi_1] = [B][C] = \begin{bmatrix} a_{11} & a_{12} & a_{13} \\ a_{21} & a_{22} & a_{23} \\ a_{31} & a_{32} & a_{33} \end{bmatrix} \qquad (20)$$

with
$$[\Phi_2] = [B], \qquad [\Psi_1] = [C]$$

then,

$$\begin{aligned}
\bar{i}_2 \cdot \bar{k}_1 &= a_{13} \\
\bar{j}_2 \cdot \bar{k}_1 &= a_{23} \\
\bar{k}_2 \cdot \bar{k}_1 &= a_{33} \\
\bar{i}_1 \cdot \bar{j}_1 &= a_{12} \\
\bar{i}_2 \cdot \bar{i}_1 &= a_{11}
\end{aligned} \qquad (21)$$

But

$$\begin{aligned}
a_{13} &= b_{11}c_{13} + b_{12}c_{23} + b_{13}c_{33} \\
a_{23} &= b_{21}c_{13} + b_{22}c_{23} + b_{23}c_{33} \\
a_{33} &= b_{31}c_{13} + b_{32}c_{23} + b_{33}c_{33} \\
a_{12} &= b_{11}c_{12} + b_{12}c_{22} + b_{13}c_{32} \\
a_{11} &= b_{11}c_{11} + b_{12}c_{21} + b_{13}b_{31}
\end{aligned} \qquad (22)$$

where

$$\begin{aligned}
b_{11} &= \bar{i}_1 \cdot \bar{I} = \cos\phi_2 \cos\theta_2 \\
b_{12} &= \bar{i}_2 \cdot \bar{J} = \sin\phi_2 \cos\theta_2 \\
b_{13} &= \bar{i}_2 \cdot \bar{K} = -\sin\theta_2 \\
b_{21} &= \bar{j}_2 \cdot \bar{I} = -\sin\phi_2 \cos\psi_2 + \cos\psi_2 \cos\phi_2 \sin\theta_2 \sin\psi_2 \\
b_{22} &= \bar{j}_2 \cdot \bar{J} = \cos\phi_2 \cos\psi_2 + \sin\phi_2 \sin\psi_2 \sin\theta_2 \\
b_{23} &= \bar{j}_2 \cdot \bar{K} = \cos\theta_2 \sin\psi_2 \\
b_{31} &= \bar{k}_2 \cdot \bar{I} = \sin\phi_2 \sin\psi_2 + \cos\phi_2 \sin\theta_2 \cos\psi_2 \\
b_{32} &= \bar{k}_2 \cdot \bar{J} = -\cos\phi_2 \sin\psi_2 + \sin\phi_2 \sin\theta_2 \cos\psi_2 \\
b_{33} &= \bar{k}_2 \cdot \bar{K} = \cos\theta_2 \cos\psi_2 \\
c_{11} &= \bar{i}_1 \cdot \bar{I} = \cos\phi_1 \cos\theta_1 \\
c_{12} &= \bar{j}_1 \cdot \bar{I} = -\sin\phi_1 \cos\psi_1 + \cos\phi_1 \sin\theta_1 \sin\theta_1 \sin\psi_1 \\
c_{13} &= \bar{k}_1 \cdot \bar{I} = \sin\phi_1 \sin\psi_1 + \cos\phi_1 \sin\theta_1 \cos\psi_1
\end{aligned}$$

$$c_{21} = \bar{i}_1 \cdot \bar{J} = \sin\phi_1 \cos\theta_1$$
$$c_{22} = \bar{j}_1 \cdot \bar{J} = \cos\phi_1 \cos\psi_1 + \sin\phi_1 \sin\theta_1 \sin\psi_1$$
$$c_{23} = \bar{k}_1 \cdot \bar{J} = -\cos\phi_1 \sin\psi_1 + \sin\phi_1 \sin\theta_1 \cos\psi_1$$
$$c_{31} = \bar{i}_1 \cdot \bar{K} = -\sin\phi_1$$
$$c_{32} = \bar{j}_1 \cdot \bar{K} = \cos\theta_1 \sin\psi_1$$
$$c_{33} = \bar{k}_1 \cdot \bar{K} = \cos\theta_1 \cos\psi_1$$

The above are all known quantities with ϕ_1, θ_1, ψ_1, ϕ_2, θ_2, ψ_2 determined.

Therefore,

$$\theta = -\sin^{-1}(a_{13}) \qquad \text{(radio-ulnar deviation)} \qquad (23)$$

$$\psi = \sin^{-1}\frac{a_{23}}{\cos\theta}, \quad \text{or}$$

$$\text{(pronation-supination)} \qquad (24)$$

$$\psi = \cos^{-1}\frac{a_{33}}{\cos\theta}$$

$$\phi = \sin^{-1}\frac{a_{12}}{\cos\theta}, \quad \text{or}$$

$$\text{(flexion-extension)} \qquad (25)$$

$$\phi = \cos^{-1}\frac{a_{11}}{\cos\theta}$$

The above equations were then programmed on a digital computer for final results. Verification tests were carried out on rigid blocks oriented at known angles. The resolution obtained was within $\pm 4^\circ$ based on the experimental and analytical methods.

RESULTS

There were a total of sixty-three subjects studied. Forty were normal with no clinical or radiologic evidence of joint disease. Twenty-three were patients with confirmed rheumatoid arthritis[11] or osteoarthritis[12]. The mean age among the normals is 36.5 with a range of 28–69 years, and the averaged age for the diseased subjects is 59.8 with a range of 45–86 years. Only the joint orientation angles in various hand functions and the maximum range of motion are presented here.

The study of normal finger joint motion indicated that metacarpo-phalangeal motion had considerable variability. In flexion, there was a range of 65-107° from neutral position when measured with respect to the metacarpals. The mean values (n = 40) of flexion for each finger are 83° for the index finger, 90° for the long finger, 88° for the ring and 90° for the little finger. In extension the range was −6 to −45° with average for index −22°; long −22°; ring −23°; and little finger −34° (negative sign designates extension).

The interphalangeal joint results are more consistent, with a smaller range of variation. The PIP joint has a mean value of flexion of 101° for the index finger, 103° for the long finger, 105° for the ring finger and 103° for the little finger. The variation among the fingers has a range of 92-125°. Extension was, as expected, limited to a range of 10-12° for all digits except the little finger which had a smaller average value of only 6.5°. The distal finger joint (DIP) had a mean flexion angle of 73° for the index finger, 80° for the long finger, 75° for the ring finger and 78° for the little finger. Extension angle averaged 11.45° from all fingers with a variation range of 10-23°.

All the above measurements were obtained based on active motion of the hand. Passive range of motion was also evaluated but the results were not conclusive for a final report. Abduction-adduction and rotational motions were generally very small for the DIP and PIP joints but they do exist due to oblique orientation of the flexion-extension axis of rotation. At the MP joints, such motions are significant. Aside from those measured under pinch and grasp functions, these out-of-plane movements have not been reduced from the raw data due to lack of time in this period of investigation. For the convenience of assessing tendon excursion in the future, total range of flexion-extension for each finger was also determined. The values of 256°, 274°, 268° and 272° were obtained for the index, long, ring and little fingers respectively.

In quantitating the three-dimensional joint orientations during hand functional activities, selected fingers and thumbs from a limited number of subjects were analyzed as an example. Only tip pinch and grasp of a 38mm diameter cylinder were evaluated (Table 4-1). More complete data on normal thumbs and patients with artificial CMC joint replacement have been completed and will be presented later in this section.

In the analysis of thumb joint motion, 27 normal subjects and 17 arthritic patients were included in the preliminary study. Eleven of the patients had rheumatoid arthritis and six had osteoarthritis, mainly involving the CMC and IP joints. No attempt was made to differentiate the data according to sex and age. Correlation with clinical disease stage was not investigated either. The results were mainly used to justify the feasibility of the present methodology. Important clinical application of such information awaits better and more complete analysis of patient series.

Table 4-1   Joint orientation angles in hand functions (degrees).

Function	Hand Unit	Joint	Orientation Angles (mean value)		
			Flex(+)/ Ext(−)	Rad. Dev (+)/ Ulnar(−)	Pron(+)/ Sup(−)
Tip Pinch	Index†	DIP	36.5	0.1	1.9
		PIP	42.1	5.2	− 5.0
		MP	45.8	11.6	− 5.9
	Long†	DIP	34.4	6.5	−16.1
		PIP	62.3	6.5	5.2
		MP	37.4	− 8.3	0.2
	Thumb*	IP	67.6	5.4	4.5
		MCP	10.1	− 0.5	16.1
		CMC	28.0	18.0	19.0
Grasp (38mm cylinder)	Thumb*	IP	60.7	3.3	4.5
		MCP	35.5	1.0	3.6
		CMC	25.0	10.0	20.0
	Index†	DIP	37.5	−11.2	4.9
		PIP	53.1	− 2.5	− 9.0
		MP	65.2	6.7	−15.6
	Long†	DIP	34.9	1.7	−15.4
		PIP	61.1	− 8.1	3.0
		MP	79.4	− 5.2	− 7.6
	Ring†	DIP	34.2	3.9	1.7
		PIP	61.1	− 6.7	− 0.2
		MP	57.0	−12.1	4.0
	Little†	DIP	29.3	6.2	3.5
		PIP	39.2	− 8.0	12.1
		MP	64.6	2.9	− 3.9

* Data based on 19 normal subjects.
† Data based on 2 normal subjects.

In the analysis of thumb joint motion, there is a large variation among the normals and the results for most arthritic patients fell within the lower third of the normal range. However, in studying flexion-extension range of motion alone, the arthritic patients were found to be well below the normal range.

Normal interphalangeal joint motion was flexion-extension $100° \pm 9$; abduction-adduction $7.5° \pm 10$; rotation $8.4° \pm 9$. The last two motion components were primarily caused by the asymmetrical articulating joint surfaces which produce an oblique axis of rotation at the IP joint. In osteoarthritis, flexion-extension averaged 82° and in rheumatoid arthristis 74°. Abduction-adduction was 9.3° in osteoarthritis (slightly increase of 2.5°) but only 3.7° in rheumatoid arthritis (a decrease of 3.8°). Rotation averaged 6.9° and 6.4° in osteoarthritis and rheumatoid arthritis, respectively. Metacarpophalangeal joint motion was flexion-extension 45° ± 16; abduction-adduction 8.7° ± 3.2; rotation 12.1° ± 4. Flexion was decreased in all arthritic patients by an average of 28° ± 15 with a mean of 19° motion in rheumatoids and 41° in osteoar-thritics. There was little difference in abduction-adduction or rotation at the MCP joint between the two types of patients.

A complete range of trapeziometacarpal (or carpometacarpal) joint motion has been unrecordable prior to this study. Previous studies describes only abduction (plane of palm or perpendicular to palm) and opposition (retro-pulsion and antepulsion) motions which are difficult to interpret and lack of baseline reference. From this study, based on well-defined definitions in rigid body motion, the following trapeziometacarpal joint motions for normals were obtained; flexion-extension of 53°, abduction-adduction of 42° and

Table 4-2   CMC joint range of motion under different functional positions. The relative motion between the first and the third metacarpals are also shown for reference.

	Thumb Functional Position	ϕ(Flex-Ext) + = Flex − = Ext	θ(Abd-Add) + = Abd − = Add	ψ(Pron-Sup) + = Pron − = Sup
	Extension	20.9 ± 8.5	23.8 ± 8.1	100.4 ± 10.9
1st MC → 3rd MC	Abduction	29.9 ± 9.8	10.4 ± 11.1	108.8 ± 10.1
	Adduction	22.9 ± 6.3	7.1 ± 7.0	107.3 ± 11.9
	Flexion	24.8 ± 5.6	− 5.6 ± 9.0	110.0 ± 10.9
	Ranges	12.2 ± 5.1	29.4 ± 9.7	14.4 ± 6.9
	Extension	7.6 ± 8.8	− 26.1 ± 8.1	33.3 ± 10.4
1st MC → Trapezium*	Abduction	24.3 ± 12.8	− 21.2 ± 8.8	31.5 ± 11.0
	Adduction	28.7 ± 9.2	− 28.1 ± 7.7	28.1 ± 11.6
	Flexion	41.2 ± 10.3	− 29.5 ± 5.6	23.9 ± 11.4
	Ranges	33.2 ± 10.2	11.6 ± 5.1	11.9 ± 5.4

*Relative angles between the third metacarpal and the trapezium are: $\phi_0 = 48°$ $\theta_0 = 38°$ $\psi_0 = 81°$

rotation of 17°. In osteoarthritic patients, flexion decreased to less than half of normal value (21°). Abduction was never achieved since the test subjects maintained an adducted position of 4–19° throughout the entire experiment. Rotation was found to have a mean value of 24°. In both pinch and grasp, there was also a loss of flexion (24° in pinch and 31° in grasp) and abduction when compared to normals.

The range of motion and joint orientation measurement technique developed was applied to study a group of patients before and after total joint replacement at the thumb CMC joint. This prosthesis, developed at Mayo, consists of a metal ball with a seating post to be cemented in the trapezium and a polyethylene socket with an extended stem to be fixed in the proximal canal of the first metacarpal. Therefore, the artificial CMC joint has a ball-and-socket configuration.

A total of 26 patients (24 women and 2 men) were studied with a mean follow-up of 22 months (ranging from 6 to 30 months). The mean age of this group was 61 years of age (49 to 71 years), within which five had rheumatoid joint disease and 21 had degenerative joint disease. The detailed range of motion data for the patient group is summarized in Table 4–2. The range of

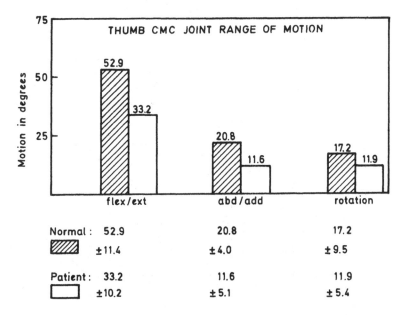

Fig. 4-11 Comparison of thumb CMC joint range of motion between the normals and patients after total joint replacement.

motion and functional strength of the thumb were significantly improved after joint replacement. When compared to the normals, more than half of the normal range of CMC joint motion was achieved after surgery (Fig. 4–11). This range of motion is considered adequate to accomplish the essential activities of daily living involving the thumb[9,10,11]. Such evaluation has helped to prove that CMC joint replacement is effective in improving the patients' hand functions, in addition to pain control.

Dexterity testing was performed by all subjects. Males were slightly faster (more coordinated) than females. The former averaging 16.7 seconds and the latter 17.6 seconds. The overall averaged timed pegboard test result was 16.9 seconds for normals. In patients with arthritis, the manipulative ability was reduced significantly with an overall result of 27.2 seconds as opposed to the normal value of 17 seconds. The rheumatoid arthritic patients were slightly worse than the osteoarthritic patients in this functional category. A comparison of these values with the CMC joint replacement patient is illustrated in Fig. 4–12. Again, CMC joint replacement had a significant effect on the improvement of the patients' prehensile manipulation abilities. A new dexterity board has been developed to expand the test criteria by including additional functions which involve the whole hand (Fig. 4–13). A baseline is

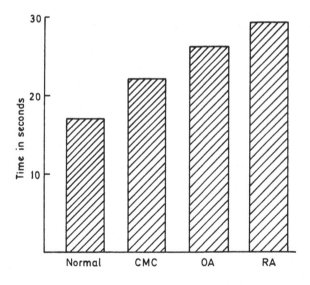

Fig. 4-12   Comparison of dexterity time in normals, patients with different diseases, and patients after CMC joint replacement.

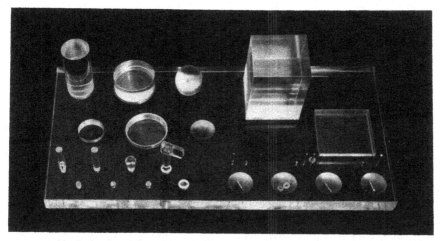

Fig. 4-13 Newly developed dexterity testing board which includes both the manipulation ability of the thumb and the fingers, as well as the adaptability of the hand in handling objects of varying sizes.

being established for this board so that patients can be evaluated with a proper reference.

DISCUSSION

To test the validity of this method, a plastic model with blocks oriented at known angles was evaluated (Fig. 4-14). "T" markers taped to the surface of each block served as reference points for the construction of the coordinate axes. Biplanar X-rays were taken of the model and the orientation angles calculated and compared to the exact known angles. The variation was found within plus and minus two degrees. Such accuracy was considered excellent for our application. In actual patient study, if one excludes the soft tissue effect which was found to be minimal, this method was considered quite adequate to determine the intricate movement occurring at the small and concealed CMC joint.

In order to test the reliability of our technique, markers of known length were placed on cubes of known dimensions (Fig. 4-15). The cubes were placed at various locations within the cassette grid field. Biplanar roentgenograms were taken of the cubes in order to determine if a change of position would alter the length of the projected markers. This method helped to determine how reproducible and how reliable the biplanar X-ray technique is.

Aside from checking known angles and distances, two additional parameters of measurement were used to verify the accuracy of the method used in patient analysis. The first parameter is the magnitude of the common perpendicular, S, as defined in the previous section. If S is small, then the error of measurement will be insignificant. Graphically, S is described as the shortest distance between the two X-ray beams which should theoretically intersect at the marker point. Because of experimental error, these beams can never intersect and the midpoint of S is used as the approximate position of the marker point (Fig. 4–6). Hence, the magnitude of S can be regarded as an indicator for possible experimental error.

The second parameter used to examine the data accuracy is the orthogonality of the coordinate axes determined. The orientation of the coordinate axes is based on the external "T" marker position. Deviations from orthogonality can occur due to large S values and the mathematical manipulations required in coordinate reconstruction. In addition, the "T" markers may not be exactly 90° due to accidental bending or poor construction. Small deviations from 90° would further reflect the accuracy of the entire experiment and theoretical procedure.

The results of the plastic model show that the angles calculated by this technique differ from the exact values by an average of six percent. The test of the

Fig. 4–14 Plexiglas cubes with "T" markers attached for verification study.

cube blocks showed that the difference between the measured locations and the calculated locations from the biplanar X-rays was on the average of only two percent. The above two tests demonstrate that the use of the present method is quite accurate in determining TMC joint orientation and movement. The maximum magnitude of the common perpendicular was 1.27 cm with a mean of only 0.38 cm for all subjects studied. In the orthogonality check, the maximum deviation from 90° was ± 4°. The average value for all subjects included in the experiment is very close to the required 90° between the coordinate axes defined. This further validates the reliability of the technique used.

The normal and patient data presented in this chapter should only be considered as preliminary results aimed to test the applicability of the method developed. Although the results can be regarded as representative values on both normal and abnormal subjects, a complete data base must be established through carefully designed studies, including a large number of normal subjects and patients, to allow adequate statistical analysis. However, the present joint orientation information can be used for joint and tendon force analysis. As far as the feasibility of the method is concerned, it has been successfully established.

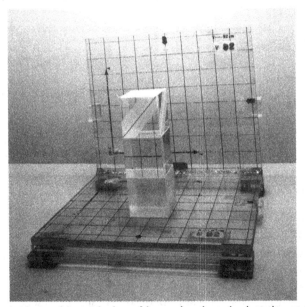

Fig. 4–15 Plastic blocks with metal wires of known lengths and orientation angles for experimental verification of the accuracy and reliability of the experimental and analytical methods used.

SUMMARY

The practical application of this investigation encompasses a major aspect of arthritis affecting the hand. Effective measuring techniques of motion, strength and dexterity will provide better assessment methods for rheumatologists, physiatrists, and hand surgeons who care for these patients. Direct application of this information to joint prosthetic design and testing will be important. Any evaluation technique would be reliable if it were accurate, easy-to-use, and reproducible. These concerns have been successfully resolved for the technique developed herein from the present study of normals and selected abnormal subjects.

REFERENCES

1. Flatt, A.E., *The Care of the Rheumatoid Hand,* edn. 2., (St. Louis, C.V. Mosby Co, 1968).
2. Leach, R.E. and Bolton, P.E., "Arthritis of the carpometacarpal joint of the thumb", *J. Bone Joint Surg.,* **50–A**: 1171 (1968).
3. Pauzat, L.D., "Osteoarthritis of the trapeziometacarpal joint", *J. Bone Joint Surg.,* **31–B**: 534 (1949).
4. Ratliff, A.H.C., "Deformities of the thumb in rheumatoid arthritis", *Hand.,* **3**: 138–143 (1971).
5. Swanson, A.B., "Disabling arthritis of the base of the thumb", *J. Bone Joint Surg.,* **54–A**: 456–471 (1972).
6. Van Wetter, P., "Biométrie et physiolgie du pouce", *Acta Orth. Belgiia,* **7**: 403 (1971).
7. Eaton, R.G. and Little, J.W., "A study of the basal joint of the thumb", *J. Bone Joint Surg.,* **51–A**: 661 (1969).
8. Morrey, B.F. and Chao, E.Y., "Passive motion of the elbow joint", *J. Bone Joint Surg.,* **58A**: 501–508 (1976).
9. Chao, E.Y., Lucca, M.J. and Cooney, W.P. III, "Kinematic analysis of normal trapeziometacarpal (TMC) joint motion", *Trans. 24th Meeting: Orthopaedic Research Society,* **3**: 165 (1978).
10. DeRoos, J.P., Chao, E.Y. and Cooney, W.P. III, "A biomechanical method for evaluation of normal and pathological thumb function", *Trans. 23rd Meeting: Orthopaedic Research Society,* **2**: 201 (1977).
11. Cooney, W.P. III, Lucca, M.J., Chao, E.Y. and Linscheid, R.L., "The kinesiology of the thumb trapeziometacarpal joint", *J. Bone Joint Surg.,* **63–A**: 1371–1381.
12. Pieron, A.P., "The mechanism of the first carpometacarpal (CMC) joint. An anatomical and mechanical analysis", *Acta Orthop. Scand, Suppl.* **148** (1973).

5

HAND FUNCTIONAL STRENGTH ASSESSMENT
AND ITS CLINICAL APPLICATION

The strength of the hand in various isometric functional activities was studied. This chapter presents the instruments used, method applied and the results obtained for forty normal subjects (18 males and 22 females). In order to eliminate anthropometric variations, an attempt was made to correlate the functional strength values with several physical measurements of the hand and forearm. Common isometric hand functions in the form of tip pinch, pulp pinch, key pinch, power grip, and also the ulnar and radial deviation forces involved in each finger were evaluated using specially designed strain gauge instruments. These evaluation results were applied to the analytical model to obtain internal finger joint and tendon forces under realistic conditions. A parallel study between normals and diseased fingers can help to quantitate the degree of functional impairment in hand patients. Similar evaluation performed before and after treatment can help to assess the effectiveness of surgical or therapeutic management.

INTRODUCTION

Although functional strength is one of the fundamental determinants in hand activities, this factor has not been given enough attention in the treatment and rehabilitation of hand patients as compared to other parameters such as pain, manipulative ability, joint alignment, and sensibility. The first report of the strength of the hand measured with a dynamometer was performed and reported by Regnier[6] nearly 180 years ago. Several recent studies of similar nature were performed using either commercially available or custom designed instruments[1,2,3,4,5,7,8,9,10,11,12,13].

A variety of instruments were designed for practical application. These devices were fabricated based on pneumatic, hydraulic, or electronic principles. The present study utilizes strain gauge instruments to obtain accurate measurement. The reduction in size and ease of calibration offer a wide range of hand functional evaluation without causing any discomfort to the subjects to be examined. The construction of the devices also allows easy adaptation to different hand deformities.

In addition to strength evaluation, manipulative ability, joint deformity and range of motion were also analyzed to provide a complete assessment of the hand function. A timed pegboard was used to examine the manipulative functions of the hand. Biplanar X-ray technique was applied to obtain joint orientation and range of motion analysis. Graphic data report forms were developed to compile the evaluation results so that the patient's test data can be communicated to the physician and a reliable performance index can be developed to document the functional progress or deterioration of the patients.

DESCRIPTION OF INSTRUMENTS

A hand functional strength testing apparatus modified from an existing design[8] is shown in Fig. 5–1. This apparatus allows pinch, grasp and lateral deviation tests and provides quantitative measurements on the fingers involved in the function. A timed pegboard consisting of four different simple tasks to test hand manipulating ability is also shown. Different forearm and wrist stabilizing splints are used to isolate the hand functions in order to obtain con-

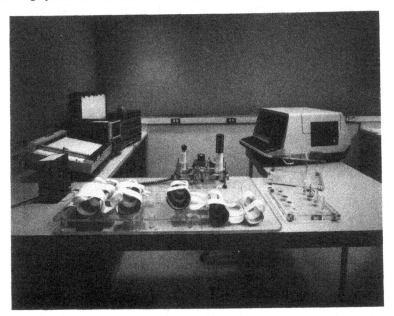

Fig. 5–1   Hand functional strength and manipulative evaluation instruments with digital force display device.

sistent and reliable evaluation data. Digital readout to register the maximum strength reading is provided for easy recording (Fig. 5-2).

The pinch meter, as shown in Fig. 5-3 is made of two metal beams separated and fastened at the base to form a tuning-fork structure. Strain gauges are

(a)

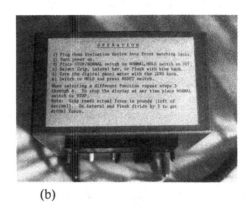

(b)

Fig. 5-2 (a) Digital force display device.
(b) Operational instructions are shown on top of the device.

Fig. 5-3 Pinch Meter.

applied on both sides of one beam at its base to provide accurate strain reading as the beam is bent by the pinch action applied at the tip of this device. A nylon tube is used to contain the pinch meter and to protect the strain gauges while the examiner can manipulate the device for convenient patient evaluation. The tip of the pinch meter has two circular disks covered with soft silastic rubber to allow pinch actions without causing discomfort to the patients. The space between the pinch surfaces of the device measures about 1cm which is similar to the regular pinch meter used by hand surgeons. As the subject pinches the padded disks, the applied forces tend to bend either beam of the device. The strain gauges record the small deflection at the base of the cantilever beam which provides a direct measurement of the applied pitch force. A careful calibration was performed so that the gauge reading can be properly scaled to provide accurate measurement of pinch strength directly in units of kg.

The function of grasp (or power grip) is measured by two different devices. The first device (Fig. 5–4) is a simple grasp meter constructed based on the same principle applied in the pinch meter except the size of the metal beams is larger in order to withstand the strong grasp force without bending the beams excessively. Plexiglass cylindrical covers are used to allow easy grasp action. In order to accommodate the variation of hand sizes, different diameters of the cylinder (3 cm, 4 cm and 5 cm) were constructed. In performing the grasp

Fig. 5–4 Grasp Meter with exchangeable cylindrical covers to accommodate various hand sizes.

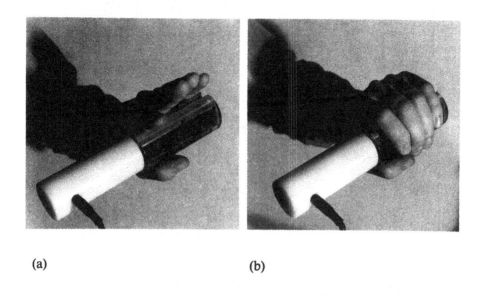

(a) (b)

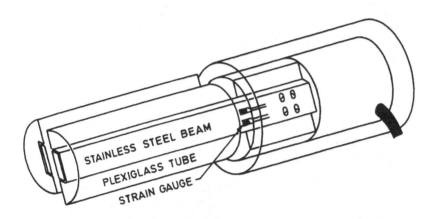

(c)

Fig. 5-5 (a) Grasp strength measurement was standardized by aligning middle phalanx of middle finger with the marker "M".

     (b) Grasp with the whole hand.

     (c) Schematic diagram of the strain-gauged grasp meter.

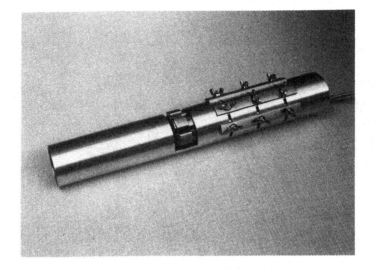

(a)

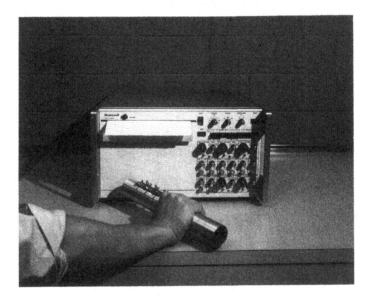

(b)

Fig. 5-6   Hand power grip meter capable of providing measurement on the compressive force produced by each phalanx of a finger.
(a)   Instrument in close-up view.
(b)   Instrument being used during evaluation.

action, standard application procedure is demonstrated in Fig. 5-5 to eliminate measurement variation. It was found important to allow the hand to completely wrap around the cylinder in a consistent manner to provide uniform readings of the strain gauges placed at the base of the meter. Proper calibration was established to provide direct reading of grasp strength in kg.

The second grasp meter, as shown in Fig. 5-6, was designed to measure isolated forces exerted by each individual phalanx of a particular finger during grasp function. Three steel beams instrumented with strain gauges at their base are fastened inside of an aluminum cylinder in a circumferential fashion. Each beam has a padded contact surface protruding from the slot of the cylinder to allow proper contact to the volar surface of the three phalanges of a finger engaged in grasp action. Each beam can be adjusted independently to provide accurate placement of the contact surface to the phalanx. Simultaneous recording of each beam allows separate reading of the strength contributed by the phalanx participating in the grasp as illustrated in Fig. 5-7. Shifting the hand longitudinally along the cylinder to allow a particular finger over the measuring slot during identical grasp can further differentiate the contribution of grasp strength by each finger.

In order to assess the intrinsic muscles of the finger, lateral deviation

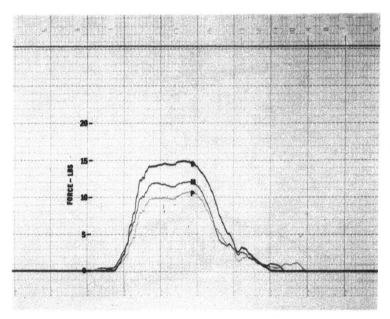

Fig. 5-7 Compressive force produced by distal phalanx (D), middle phalanx (M), and proximal phalanx (P) of long finger performing grasp.

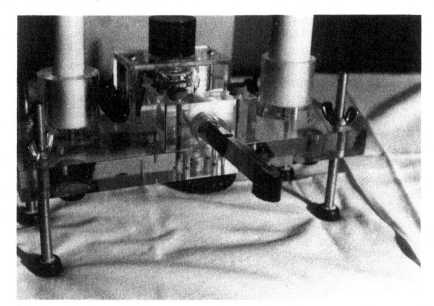

Fig. 5–8   Device for the strength measurement of lateral deviation.

Fig. 5–9   New device for the hand strength measurement. The distance between two beams can be adjusted for various functional conditions.

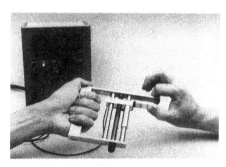
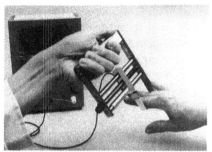

(a) (b)

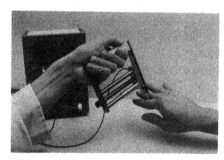

(c) (d)

Fig. 5-10 Universal functional strength tester used in different hand activities.
(a) Key pinch with varying gap.
(b) Abduction strength between index and middle fingers.
(c) Adduction strength between index and middle fingers.
(d) Opposition strength of the thumb.

strength is measured by another cantilever beam structure shown in Fig. 5-8. Strain gauges are also mounted at its base. The top of the beam is shaped in an ellipsoid form padded with soft material. When the finger is placed laterally with its DIP or PIP joint pressed against the lateral surface of the beam tip, deviation strength produced by the intrinsic muscles can be accurately registered by the strain gauges under beam bending. The beam is placed in a vertical manner to allow its narrower portion to be inserted between fingers at the PIP joint level for such measurement. Both medial and lateral deviations can be evaluated by pushing the beam to the left or right side of the examiner. The entire apparatus can be moved freely and its height can also be adjusted to best suit the patient's finger when the hand is badly deformed by disease.

Finally, a universal gauge is designed to provide strength measurement on other forms of finger function. As shown in Fig. 5-9, this gauge allows one of the two cantilever beams to be adjusted so that the distance between them can be changed to adapt to a particular functional test. The base of the adjustable beam is instrumented to allow direct reading of the applied strength. The surfaces of both beams are padded with soft material to avoid discomfort during evaluation. This universal gauge can be utilized to measure functional strength of the fingers and thumb under various conditions including pinch of different object sizes, grasp, abduction, adduction, tapping, opposition, etc. A few examples on the use of this gauge are illustrated in Fig. 5-10. A similar readout meter (Fig. 5-2) is used to provide instantaneous test results in digital form.

EXPERIMENTAL METHOD

Forty normal subjects were studied in order to evaluate the feasibility of the instruments developed here and the test data variation. Several abnormal patient groups were also examined to test the sensitivity and adaptability of the instruments to hands with joint deformity and substantial loss of strength. Four basic functional activities as shown in Fig. 5-11 and the deviation strength of each finger were performed.

Tip pinch measures the compressive strength between the tips of a finger and the thumb. The pulp pinch exerts the compression through the pulps of the distal phalanx. The lateral pinch (or key pinch) is applied through the lateral aspect of the DIP joint of a finger without ulnar support of the rest of the fingers. If the lateral pinch is exerted through the index finger, stabilized by the remaining fingers, it is defined as the chuck pinch.

For overall grasp strength, the simple grasp meter shown in Fig. 5-4 was used. In order to determine the contribution of each finger to grasp strength and to record the individual phalangeal force, the special grip meter as seen in Fig. 5-6 will be applied. The radial and ulnar deviation forces are measured by

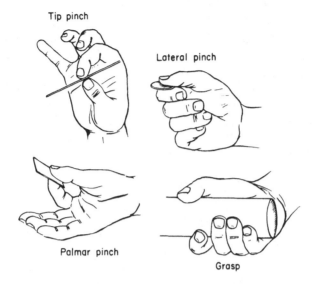

Fig. 5-11 Various isometric functions of hand.

the simple device shown in Fig. 5-8 rather than the universal gauge since the application of this more complex instrument awaits further experimental study.

In normal subjects, the predominant hand was evaluated by placing it in the splint to eliminate the effect of the wrist position. In order to prevent muscle fatigue, the subject was instructed to relax after each strong exertion of force. Three measurements were obtained for each type of function and the mean value was used to accommodate variation. These measurements were also correlated with different physical dimensions of the hand to establish certain scale factors for the purpose of normalization.

The strength measurements plus all other pertinent clinical and physical data of a subject are recorded in the evaluation forms. The joint angular orientation is analyzed by the biplanar X-ray technique described in the previous chapter. The results of functional forces and joint orientation angles are applied to the analytical model derived before in order to calculate the realistic tendon and joint forces among normal and patient populations. These results will be used to quantitate the pattern and magnitude of functional impairment in diseased and disabled hand patients. Such evaluation is also utilized to assess the functional improvement after surgical reconstructive procedures.

RESULTS OF NORMAL SUBJECTS

Based on the instruments and procedure described above, forty normal subjects (18 males and 22 females) were evaluated. Normalizing hand functional strength among the normal population has been a difficult problem due to its dependence on many factors and the large variation existing among any population profile to be studied[15,16,17,18]. We also attempted to normalize this data based on anthropometric measurements. Table 5-1 summarizes the ranges, averages, and standard deviations of these measurements for different functions studied. These results are further divided according to sex. The male group is consistently stronger than the female group. In order to eliminate

Table 5-1   Strength of normal hand (kg)

		MALE (n = 18)				FEMALE (n = 22)			
		MEAN	S.D.	MAX.	MIN.	MEAN	S.D.	MAX.	MIN.
Radial	Index	4.43	1.45	7.12	1.61	3.34	0.85	5.45	1.67
Lateral	Long	4.43	1.76	8.18	1.82	2.89	0.79	4.73	1.78
Movement	Ring	3.42	1.79	7.61	1.67	2.35	0.84	3.82	1.09
at PIP	Little	2.58	1.59	7.52	1.33	1.82	0.92	3.55	0.10
Ulnar	Index	4.57	1.81	7.21	1.64	3.29	0.90	5.76	1.94
Lateral	Long	4.75	1.99	9.58	1.64	2.79	0.95	5.18	1.09
Movement	Ring	3.63	1.83	9.20	1.48	2.35	0.94	4.36	1.24
at PIP	Little	2.66	0.93	5.15	1.45	2.11	0.76	3.70	1.03
Radial	Index	7.65	1.64	10.79	4.36	6.03	1.21	8.15	3.76
Pinch	Long	6.95	1.55	9.14	4.21	5.17	0.98	0.85	2.82
Tip	Index	6.43	1.01	9.70	5.33	4.82	1.08	6.94	2.48
Pinch	Long	6.46	1.94	10.20	3.97	4.66	1.24	7.33	2.79
Pulp	Index	6.58	1.35	8.45	4.45	4.55	0.92	6.97	3.06
Pinch	Long	6.36	1.48	8.61	3.06	4.57	0.92	7.00	3.30
Grasp		37.51	8.69	58.48	23.20	22.30	6.28	36.70	11.97

anthropometric variation, several physical parameters were also measured and correlated with the results. These physical parameters are:

(1) hand length (cm)
(2) forearm circumference (proximal ⅓, cm)
(3) palm circumference (cm)
(4) thenar eminence (cm)
(5) hypothenar eminence (cm)
(6) body weight, and
(7) age (years).

If any one of the anthropometric parameters can provide close correlation with the hand functional strength, it will be selected as the common denominator to normalize the data. Tables 5-2 and 5-3 list the linear correlation coefficients between functional strength and these physical parameters for both the male and the female groups. Apparently, no close correlation can be found based on the linear model. Additional study with a larger sample size and range is indicated if a reliable scale factor is to be established.

Table 5-2 Linear correlation coefficient of hand strength (male, $n = 18$)

		HAND LENGTH (cm)	FOREARM CIRCUM-FERENCE (cm)	PALM CIRCUM-FERENCE (cm)	THENAR EMINENCE (cm)	HYPO-THENAR EMINENCE (cm)	BODY WEIGHT (lbs)	AGE (years)
Radial	Index	.46	.12	− .03	− .16	.15	.25	− .21
Lateral	Long	.61	.08	.06	− .13	.18	.16	− .20
Movement	Ring	.63	.08	.13	− .16	.09	.12	− .26
at PIP	Little	.41	− .15	− .07	− .24	.11	− .14	− .44
Ulnar	Index	.43	.11	.01	− .26	.20	.30	− .11
Lateral	Long	.60	.16	.02	− .03	− .02	.14	− .14
Movement	Ring	.51	− .14	− .19	− .13	− .11	.04	− .29
at PIP	Little	.28	− .27	− .31	− .20	.06	.37	.07
Radial	Index	.48	.06	− .13	.04	− .05	.04	− .33
Pinch	Long	.28	.40	− .20	.57	− .33	.06	− .27
Tip	Index	.19	.02	− .28	.25	.12	.05	− .33
Pinch	Long	.62	.24	.04	.14	− .28	.14	− .33
Pulp	Index	.32	.37	− .02	.42	− .13	.08	− .31
Pinch	Long	.59	.47	.17	.32	− .18	− .12	− .41
Grasp		.25	.30	.00	.20	− .26	− .19	− .50

Table 5-3   Linear correlation coefficient of hand strength (Female, $n = 22$).

		HAND LENGTH (cm)	FOREARM CIRCUM- FERENCE (cm)	PALM CIRCUM- FERENCE (cm)	THENAR EMINENCE (cm)	HYPO- THENAR EMINENCE (cm)	BODY WEIGHT (lbs)	AGE (years)
Radial	Index	.16	.23	− .09	− .04	− .23	.26	.16
Lateral	Long	.04	.38	.14	.11	− .12	.27	.12
Movement	Ring	− .03	.41	− .21	.08	− .20	.37	.02
at PIP	Little	− .07	.36	− .11	− .06	− .28	.29	.16
Ulnar	Index	− .03	.09	− .32	− .04	− .06	.23	− .01
Lateral	Long	.04	.26	− .04	.00	− .04	.22	− .02
Movement	Ring	.14	.30	− .00	− .02	− .05	.30	.02
at PIP	Little	.31	.25	.12	.22	.05	.31	.13
Radial	Index	.26	.28	.50	.11	− .11	.36	.08
Pinch	Long	.27	.29	.48	.17	− .01	.47	.17
Tip	Index	.22	.52	.27	.36	− .06	.54	.14
Pinch	Long	.40	.46	.55	.37	.13	.38	.03
Pulp	Index	.36	.20	.44	.02	.01	.45	.08
Pinch	Long	.56	.39	.42	.22	.29	.36	.26
Grasp		.53	.45	.12	.12	.19	.37	− .08

The force exerted by each phalanx during grasp is listed in Tables 5-4 through 5-7 for each finger. In these tables, all forty subjects were used to compile the data. Six different types of phalangeal force distribution previously classified were used to group the results[19]. In Type 1, the distal phalanx carries the largest force followed by the middle phalanx and then the proximal phalanx, Type 2 is similar to Type 1 except the force distribution between the middle and the proximal phalanges is reversed. In Types 3 and 4, the middle phalanx exerts the most force while the distal and proximal phalanges have reversed magnitudes. Finally, in Types 5 and 6, the proximal phalanx is dominant in force exertion followed by either the distal phalanx (Type 5) or the middle phalanx (Type 6). The results show that different fingers have different dominating types of phalangeal force distribution. For some fingers, certain grasp types were not observed among the normal subjects. Consequently, for index, long, and ring fingers, there were five various types observed but for the little finger, there were only four types. In general, either the first or the second type of phalangeal force distribution was found to be

Table 5-4 Distribution of grip forces among phalanges of index finger (Pounds) ($n = 40$)

TYPE	NO. OF SUBJECTS	DISTAL PHALANX				MIDDLE PHALANX				PROXIMAL PHALANX			
		MEAN	S.D.	MAX.	MIN.	MEAN	S.D.	MAX.	MIN.	MEAN	S.D.	MAX.	MIN.
1	6	13.9	4.1	19.0	8.0	10.2	2.9	14.3	6.0	6.7	4.0	13.2	2.0
2	19	14.1	4.5	24.5	8.5	4.9	2.5	8.5	1.5	9.4	3.0	16.5	5.0
3	2	8.5	1.4	9.4	7.5	10.0	1.4	11.0	9.0	2.5	0.7	3.0	2.0
4	0	-	-	-	-	-	-	-	-	-	-	-	-
5	10	11.6	4.3	18.3	5.4	5.8	3.0	11.5	2.0	13.3	4.6	20.7	7.0
6	3	8.0	3.5	12.0	5.5	9.2	4.3	14.0	6.0	11.0	2.7	14.0	9.0

Table 5-5 Distribution of grip forces among phalanges of middle finger (Pounds) ($n = 40$)

TYPE	NO. OF SUBJECTS	DISTAL PHALANX				MIDDLE PHALANX				PROXIMAL PHALANX			
		MEAN	S.D.	MAX.	MIN.	MEAN	S.D.	MAX.	MIN.	MEAN	S.D.	MAX.	MIN.
1	23	15.3	5.9	29.0	7.5	9.2	3.6	15.5	3.0	5.5	2.7	12.3	1.0
2	13	15.3	4.8	25.0	10.0	6.0	4.0	16.0	2.5	8.2	3.3	16.0	4.5
3	1	13.5	0.0	13.5	13.5	15.0	0.0	15.0	15.0	10.0	0.0	10.0	10.0
4	1	9.0	0.0	9.0	9.0	15.0	0.0	15.0	15.0	11.5	0.0	11.5	11.5
5	2	14.2	6.2	18.8	10.0	8.3	2.9	10.3	6.3	15.2	7.3	20.3	10.0
6	0	-	-	-	-	-	-	-	-	-	-	-	-

Table 5-6   Distribution of grip forces among phalanges of ring finger (Pounds) ($n = 40$)

TYPE	NO. OF SUBJECTS	DISTAL PHALANX				MIDDLE PHALANX				PROXIMAL PHALANX			
		MEAN	S.D.	MAX.	MIN.	MEAN	S.D.	MAX.	MIN.	MEAN	S.D.	MAX.	MIN.
1	25	9.9	2.7	15.3	3.5	6.2	2.8	12.0	1.5	3.3	2.4	10.5	0.5
2	7	8.7	3.5	14.5	5.5	4.5	2.5	8.5	2.0	5.6	2.9	10.0	3.0
3	5	7.9	2.1	11.6	6.5	10.5	3.3	15.0	7.5	3.3	4.1	10.6	0.8
4	1	6.0	0.0	6.0	6.0	7.5	0.0	7.5	7.5	6.5	0.0	6.5	6.5
5	2	8.4	1.3	9.4	7.5	7.5	1.4	8.5	6.5	9.6	0.2	9.8	9.5
6	0	-	-	-	-	-	-	-	-	-	-	-	-

Table 5-7   Distribution of grip forces among phalanges of little finger (Pounds) ($n = 40$)

TYPE	NO. OF SUBJECTS	DISTAL PHALANX				MIDDLE PHALANX				PROXIMAL PHALANX			
		MEAN	S.D.	MAX.	MIN.	MEAN	S.D.	MAX.	MIN.	MEAN	S.D.	MAX.	MIN.
1	27	6.9	2.4	11.0	3.5	4.5	2.3	10.5	1.0	1.6	1.2	5.0	0.5
2	9	6.9	2.9	12.5	4.0	2.2	1.6	4.2	0.5	3.7	2.1	7.5	0.5
3	1	4.0	0.0	4.0	4.0	4.0	0.0	4.0	4.0	1.0	0.0	1.0	1.0
4	0	-	-	-	-	-	-	-	-	-	-	-	-
5	3	6.2	2.0	8.0	4.0	1.3	0.6	2.0	0.8	7.5	2.2	9.0	5.0
6	0	-	-	-	-	-	-	-	-	-	-	-	-

most frequent for all fingers. Based on these results, the proportion of grip force distributions among the phalanges was summarized in Table 5-8. The numbers were normalized with respect to the strength at the distal phalanx. These results reflect the great variability in grasp force distribution which will definitely influence the associated tendon and joint forces to be calculated in the normative hand model.

Table 5-8 Proportion of grip force distribution among phalanges. (total of 40 subjects)

TYPE	NO. OF SUBJECTS	DISTAL PHALANX	MIDDLE PHALANX	PROXIMAL PHALANX
I. INDEX				
1	6	1.0	0.74	0.43
2	19	1.0	0.34	0.66
3	2	1.0	1.20	0.30
4	0	1.0	-	-
5	10	1.0	0.47	1.16
6	3	1.0	1.14	1.43
II. LONG				
1	23	1.0	0.59	0.35
2	13	1.0	0.36	0.53
3	1	1.0	1.11	0.74
4	1	1.0	1.68	1.28
5	2	1.0	0.58	1.05
6	0	1.0	-	-
III. RING				
1	25	1.0	0.57	0.28
2	7	1.0	0.48	0.60
3	5	1.0	1.32	0.32
4	1	1.0	1.27	1.07
5	2	1.0	0.88	1.15
6	0	1.0	-	-
IV. LITTLE				
1	27	1.0	0.60	0.22
2	9	1.0	0.26	0.49
3	1	1.0	1.00	0.25
4	0	1.0	-	-
5	3	1.0	0.22	1.22
6	0	1.0	-	-

The abduction-adduction strength between fingers was studied in 46 additional normal subjects (mean age of 31 years, ranging from 21–58 years) using the universal testing device (Fig. 5-9). The mean and standard deviation of this data divided according to sex is shown in Table 5-9.

Table 5-9 Abduction and adduction strength (kg)

		Abduction			Adduction		
	Fingers	23	34	45	23	34	45
Mean	Male	4.68	3.85	4.40	3.66	2.63	1.75
(S. D.)	(n = 22)	(1.63)	(0.73)	(1.09)	(0.67)	(0.54)	(0.40)
	Female	3.39	2.92	2.88	2.74	1.84	1.09
	(n = 24)	(1.06)	(0.91)	(0.90)	(0.67)	(0.48)	(0.41)

In general, abduction is stronger than adduction and comparatively, the strength is highest between the index and middle fingers. The male subjects had strengths 30–40% higher than the females. There was no statistical difference between dominant and nondominant hands. In order to correlate abduction/adduction strength with muscle size, the present data is plotted against the PCSA of the corresponding intrinsic muscles described in Chap. 2 (Fig. 5-12). Good correlation was found although the muscle moment arms to

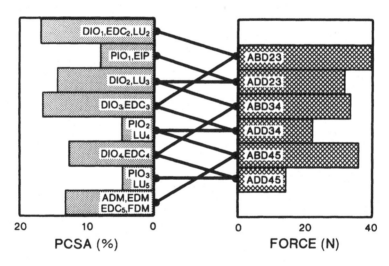

Fig. 5-12 Correlation between intrinsic muscle PCSA and functional test strength in abduction-adduction between fingers.

the MP joint were not considered. The large abduction strength between the ring and little fingers is attributed to the large muscle mass of the little finger and the strong ligament of the ring finger attached to the rest of the hand.

In order to increase the present data base, more normal subjects with a wider age range were recruited. In addition to the regular hand strength of pinch and grasp, radial and ulnar deviation forces of each finger and thumb were also measured. Table 5-10 summarizes the expanded results. Large variations still occurred in certain strength categories but these data should provide useful baseline information. It is important to recognize that the functional measurement for a particular patient can serve as their own reference, thereby, variation within a group of subjects should not matter for such application.

Table 5-10   Extended hand function strength as divided by sex ($n = 124$, 65 men and 59 women)

Type of function	Functional Strength (kg)	
	Male	Female
Grasp (n = 60)	40 ± 9	23 ± 7
Tip pinch (n = 124)	6 ± 1	5 ± 1
Pulp pinch (n = 60)	6 ± 1	5 ± 1
Key pinch (n = 84)	11 ± 2	8 ± 1
Rad. dev. (n = 60)		
Index	4 ± 1	3 ± 1
Middle	4 ± 2	3 ± 1
Ring	3 ± 1	2 ± 1
Little	2 ± 1	2 ± 1
Uln. dev. (n = 60)		
Index	4 ± 1	3 ± 1
Middle	4 ± 2	3 ± 1
Ring	3 ± 2	2 ± 1
Little	3 ± 1	2 ± 1
Thumb abd. (n = 47)	4 ± 1	3 ± 1
Thumb add. (n = 47)	7 ± 3	5 ± 2

Hand dexterity testing was performed on a large population including an older age group using the new peg board described in the previous chapter. For the sex and hand dominance effects, fifty normal (26 women and 24 men with a mean age of 34 years, ranging from 20 to 59 years) were studied. The manipulative ability as measured by the peg board time was slightly higher in women, but the difference was not statistically significant. However, the difference between the dominant and nondominant hands in both groups was statistically significant based on the paired student t-test. The most interesting finding was that when the data was divided according to age (Table 5–11), a significant increase of peg board time was found with increasing age. Although the number in the elderly age group was small, such finding seems to correspond with the general observation that age has an effect on neuro-muscular reflexes which in turn would affect the prehensile manipulation of the hand. The present method of testing appears to be effective in quantitating such physiologic change in normal subjects.

Table 5-11 Hand dexterity test results using objects of different sizes (the new peg board) and divided according to age and hand dominancy (dominant - D, nondominant -N)

Age range	Small object		Med. object		Large object		Total	
	D	N	D	N	D	N	D	N
20 - 29 (n = 22)	25 ± 3	29 ± 3	13 ± 2	14 ± 2	15 ± 1	16 ± 2	52 ± 5	60 ± 6
30 - 39 (n = 15)	26 ± 3	28 ± 4	14 ± 2	16 ± 3	15 ± 2	17 ± 2	54 ± 5	61 ± 7
40 - 49 (n = 7)	27 ± 3	36 ± 4	15 ± 2	17 ± 2	16 ± 1	18 ± 3	57 ± 4	65 ± 7
50 - 59 (n = 7)	26 ± 2	31 ± 4	14 ± 3	16 ± 3	16 ± 3	18 ± 3	54 ± 6	63 ± 9
70 - 79 (n = 4)	37 ± 5	40 ± 7	20 ± 1	22 ± 2	23 ± 4	25 ± 3	80 ± 8	87 ± 11
80 - 89 (n = 6)	47 ± 10	58 ± 18	25 ± 7	28 ± 4	27 ± 6	31 ± 8	99 ± 21	116 ± 29

Dexterity Test Time (Sec.)

FUNCTIONAL RESULTS IN PATIENT GROUPS

Four groups of patients were studied based on the testing method established. These included: 1) preoperative patients with arthritic diseases in the hand, 2) postoperative CMC joint replacement patients, 3) scleroderma patients with substantial involvement of the hands, and 4) patients with peripheral nerve neuropathy that affected normal hand function. Although these patient series were not conclusive, interesting findings have suggested many possibilities for using the present scheme in clinical applications.

A. Preoperative with Arthritic Diseases

Seventeen patients with arthritic diseases (11 RA and 6 OA) were evaluated before reconstructive procedures as a group. Their functional strength, joint range of motion, a dexterity time (based on the old peg board design) are summarized in Table 5-12. In general, the advanced arthritic patients had hand strengths only about 30% of normal, with the RA patients being weaker than the OA patients. The range of motion was also reduced significantly, especially in the RA patients. On the average, the peg board test time was prolonged to 61% above the normal value. Again, the OA patients were rated closer to the normals.

Table 5-12 Mean functional results of
normal and diseased thumbs

Thumb Functional Performance	Normal (n = 47)	Rheumatoid (n = 11)	Degenerative (n = 6)
Tip pinch (kg)	5.29	1.67	1.62
Lat. pinch (kg)	4.62	1.64	1.81
Key pinch (kg)	8.26	2.82	2.41
Chuck pinch (kg)	8.08	3.34	2.63
Abd. strength (kg)	2.65	0.75	0.67
Add. strength (kg)	4.80	1.74	1.38
Flex. strength (kg)	7.43	2.85	2.06
Ext. strength (kg)	1.83	0.62	0.63
Peg board time (sec)	16.9	28.7	25.7
IP Joint Flex.-Ext. range	106.84°	74.27°	82.55°
MP Joint Flex.-Ext. range	45.77°	19.72°	43.83°

B. *Postoperative CMC Joint Replacement*

Twenty-six patients with CMC joint replacements were studied one to two years after surgery. These patients' strength, dexterity, and range of motion were all about 60% of normal, whereas, the patients with CMC joint disease before surgery could only reach to 30–35% of the normal value (Figs. 5-13, 5-14). In addition, a group of rheumatoid patients before MP joint replacement was evaluated. It was found that their functional results were comparable to those with silastic prostheses[20]. At one year follow-up, finger joint deformities were corrected but the total arc of motion at the MP joint did not improve. Their functional strength was only 25% of the normal range. However, the joint deformity correction was found to be significant in improving patients' daily activities.

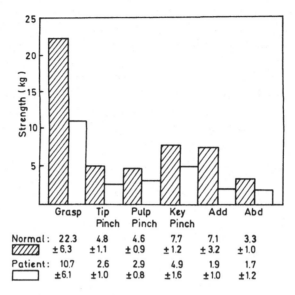

Fig. 5-13 Comparison of hand strength between normals and patients after thumb CMC joint replacement ($n = 30$).

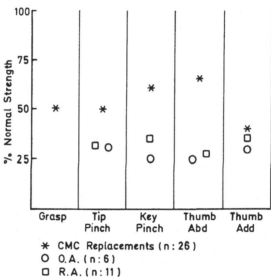

Fig. 5-14 Comparison of hand functional strength of CMC joint replacement patients ($n = 26$) and arthritic patients ($n = 17$) in percentage of normal.

C. Scleroderma

A group of 45 patients with scleroderma was studied under the general hand evaluation regime to assess the efficacy of drug and physcial therapies. An additional evaluation criterion involving skin compliance test to quantitate tissue change due to the progressive effect of the disease was performed. A surface hardness durometer was used to obtain quantitative measurement of skin compliance by a numeric reading (Fig. 5-15). Three performance indices were developed to rate the motion, strength, and skin compliance using the following formula.

$$Ip = \frac{1}{n} \sum_{i=1}^{m} \left(W_i \frac{X_i - \bar{X}_i}{\sigma_i} \right)$$

$$n = \sum_{i=1}^{m} W_i$$

where

m = number of parameters in each evaluation criterion,

X_i = patient's measured ith parameter value,

$\bar{X}_i$ = mean value of the ith parameter in normals, and

σ_i = standard deviation of ith parameter in normals.

These three index values can be properly averaged to give an overall rating of the patients before and after therapeutic and/or drug treatment. It was found that among all laboratory tests performed on these patients, the biomechanical functional evaluation results correlated the closest with the physicians' clinical examination concerning the patients' disease stage. This test procedure is now being recommended by the rheumatologists as a routine examination after the scleroderma patients undergo experimental treatment. The skin compliance has also proven to be effective, simple, and sensitive. It is now under experimental study to see if it could be used to assess joint swelling and synovitis.

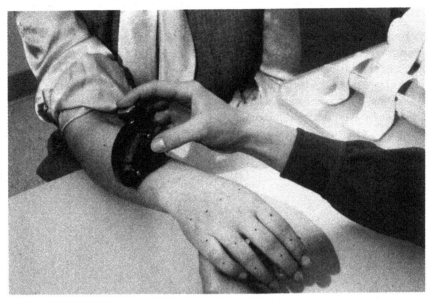

Fig. 5-15 Shore durometer used for testing of skin compliance at points marked over the dorsal surface of the hand and forearm.

D. Peripheral Neuropathy

Hand strength evaluation was performed on 32 patients with documented (by EMG examination and manual physical testing) single nerve lesions of the radial, median and ulnar nerves. The objectives of this study are two-fold. First, to document the effects of loss of individual muscles (or groups of muscles) on functional hand strength, as well as isolated finger and thumb strength, in order to have indirect verification of theoretical muscle force calculations, and secondly, to test if a procedure could be developed, based on patterns of strength distribution, to differentiate the types and levels of peripheral nerve lesions equal to or better than electromyography.

The strength of functional activities, such as grasp, pinch and abduction-adduction between fingers, was measured using the grasp and pinch meters. In addition, isolated strength such as flexion/extension and radial/ulnar deviation of the thumb and fingers was measured. The mean values of the functional strengths for the three different neuropathic populations are summarized in Table 5-13.

The absolute strengths vary with subjects. In order to facilitate the comparison of the strength for each neuropathic group, a normalization procedure

Table 5-13 Functional strength (kg) of neuropathic patients.

	Normal (n = 46)	Nerve Lesion		
		Ulnar (n = 14)	Median (n = 13)	Radial (n = 5)
Grasp	34.6	18.5	20.1	14.0
Tip Pinch	5.6	2.8	4.2	3.2
Pulp Pinch	5.1	2.4	4.2	4.0
Key Pinch	9.8	4.9	7.6	7.1
Abduction				
I/M	4.0	.97	2.6	1.0
M/R	3.1	.76	1.7	.77
R/L	3.2	.75	2.0	.91
Adduction				
I/M	3.2	1.3	2.2	2.4
M/R	2.2	.97	1.9	1.4
R/L	1.5	.65	1.2	1.5

I = index finger, M = middle finger, R = ring finger, L = little finger.

was developed. This procedure was developed based on the assumption that under normal conditions, all of the hand muscles and thus the functional strengths grow uniformly. In other words, for normal subjects there is a specific pattern of distribution of strengths among various functions. For a selected group of functions, the strengths of neuropathic patients were normalized by using least-square minimization in comparison with those of the averaged normal strengths.

Minimizing

$$E = \sum_{i=1}^{N} (\mu_i - ca_i)^2$$

Gives

$$c = \left(\sum_{i=1}^{N} \mu_i a_i \right) \bigg/ \left(\sum_{i=1}^{N} a_i^2 \right)$$

where μ_i and a_i represent the strengths of mean normal and that of patients, respectively, of a selected function. The constant c, thus defined, is then used to scale the patient strength value to the normal pattern.

A typical example which uses the strengths of the index finger in flexion, extension, radial and ulnar deviation is shown in Fig. 5-16, which shows the

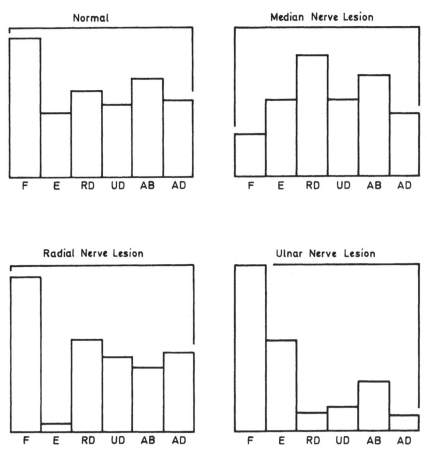

Fig. 5-16 Pattern of normalized index finger strength of normals and peripheral neuropathic patients. F = flexion, E = extension, RD = radial deviation, UD = ulnar deviation, AB = abduction and AD = adduction.

characteristic patterns of each neuropathic group. Optimal selection of the strength measurements to be included in the discrimination procedure is under investigation such that a reliable method can be developed to differentiate not only the type but also the level of the peripheral nerve lesion.

REPORT FORMS

Patient functional evaluation report forms developed have proven to be most useful for the clinicians. A two-page graphic diagram and tabulation generated by an on-line computer summarizes the patients' current and past results as related to normal ranges and is ready immediately after evaluation (Fig. 5–17). These forms will remain in the patients' histories to assist clinical evaluation. The basic format will be simplified to facilitate easy reading. After this form is optimized, a clinical trial period will be conducted to justify the evaluation as an acceptable clinical procedure for studying treatment effects and to quantitate disability levels.

DISCUSSION

Both Swanson (1970) and Dickson *et al.* (1972) stated that for functional strength there is no significant difference between the dominant and nondominant hands. Therefore, in our preliminary study of normal subjects, only the dominant hand was measured. This observation may not be true as more sensitive instruments are applied. More systematic analyses must be performed in the future in order to substantiate this finding. The current number of subjects still may not be sufficient to establish the data base, particularly in the older age group; with a larger sample size, the results can be grouped more effectively according to sex, age, physical condition and occupation.

The present measuring instruments, designed based on the strain gauge principle, provided several advantages in accuracy, ability to provide a wide range of measurement, and ease of adaptation to patients. The digital strain indicator is particularly effective in calibration and is able to record the maximum strength instantaneously. Thus, the subject can exert the maximum strength and release right away to avoid any fatigue effect. This helps the evaluator to conduct the experiment conveniently while obtaining reliable results. In addition to these functional strengths, several physical parameters such as manipulation ability, joint range of motion, and skin compliance were measured in order to provide a complete assessment on the functional status of a patient. The development of a data report form has been a significant step forward which provides a useful presentation of test results to the physicians.

The results for pulp pinch and lateral (key) pinch compare favourably with those observed by Swanson[7]. Slightly larger values were obtained in our series

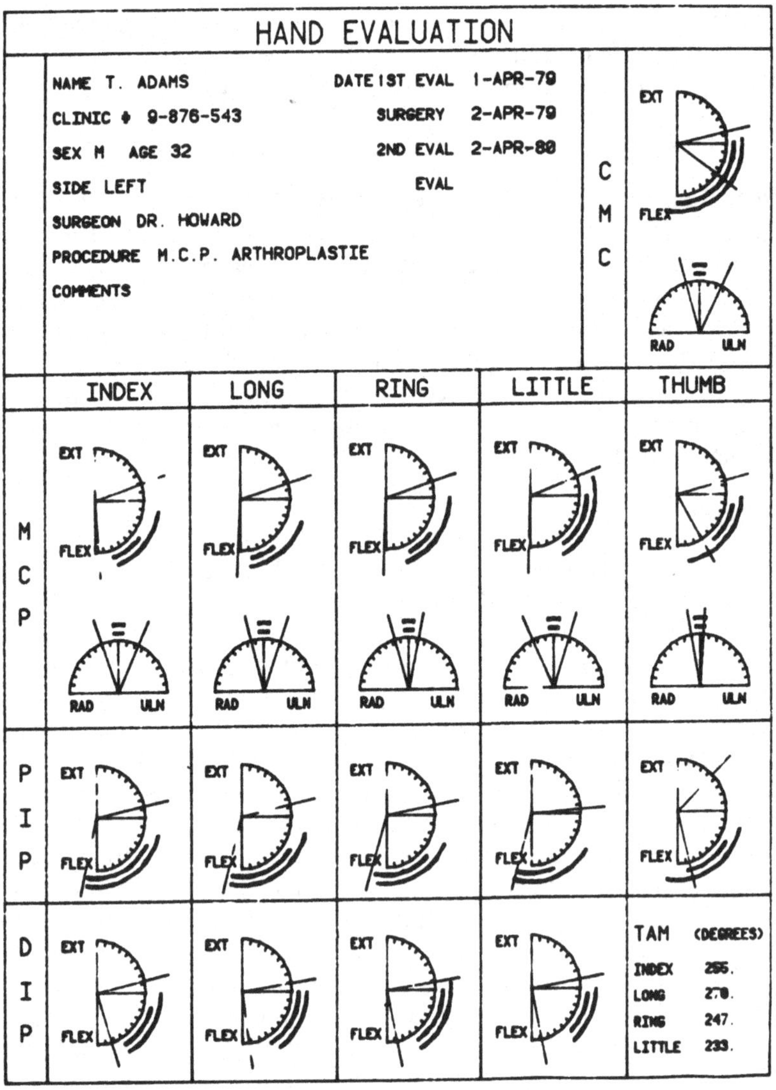

(a)

Fig. 5-17 Patient functional evaluation report form available to the clinician the same day the hand functions examination is performed.
(a) Range of motion data,
(b) Joint strength and dexterity data.

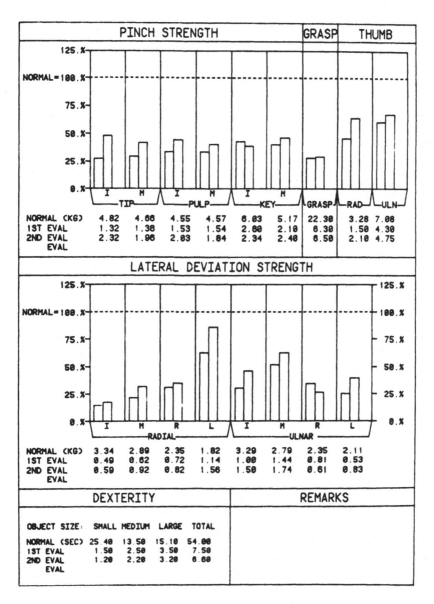

(b)

and were credited to better measuring devices and careful subject control. In the present data, the lateral pinch against the index finger is the strongest as compared to all other fingers for both the male and the female subjects. In pulp and tip pinch, the strengths generated by the long and index fingers are nearly equal for both sexes. Lateral pinch was found to be the strongest pinch action investigated (excluding the chuck pinch which is specially designed to measure the strength of the thumb).

The average power grip or grasp force is slightly lower in our measurements than that found by Swanson[7]. Our grip meter is a strain gauge instrumented cylinder with a 1.25 inch diameter which is different from the Jamar hydraulic dynamometer used by Swanson both in physical shape and measurement response. Strictly speaking, these two grip meters require two different forms of grasp, thus the strength reading will vary. The measurement by the Jamar dynamometer represents a different gripping function, and its strength cannot be directly compared with the cylindrical grasp meter presently applied.

Variations in the functional strength were correlated with the selected physical dimensions of a hand. The linear correlation coefficients obtained (Table 5-2) failed to show any significant relationship. However, the hand functional strength of the male population seems to show slightly positive correlation with the hand length and negative correlation with the age. In the female population studied, forearm circumference and body weight show a small positive correlation for all the hand strengths. Palm circumference has demonstrated a slightly positive correlation with the three basic types of pinch among female subjects. The palm circumference and forearm circumference were selected as the two key parameters in a current study since they can be used to describe indirectly the size of the intrinsic and extrinsic muscles in the hand. Since the correlation may not be linear, the nonlinear and multivariable correlation is suggested for future study.

The results of force distribution among phalanges in grasp have shown that the concept of even distribution of grasp forces on each finger is incorrect. There are six possible types of distribution of each finger according to the order of magnitude on distal, middle, and proximal phalanges. In general, the distal phalanx contributes the most to the grasp strength followed by either the middle phalanx or the proximal phalanx. The proportional ratios as seen in Table 5-8 are highly variable depending upon hand size and neuromuscular control. Therefore, in calculating tendon and joint force distribution, a large variety of phalangeal force distribution must be considered in order to provide a more complete picture of grasp action.

The results on ulnar and radial deviation strength are different from those observed by Matheson *et al.*[5]. A comparison of these results is presented in Fig. 5-18. In the male group, the current data shows that both long and index

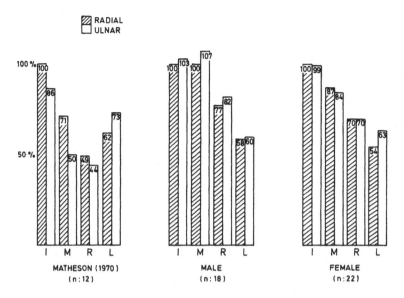

Fig. 5-18   Comparison of relative strengths of lateral movement of index (I), middle (M), ring (R), and little (L).

fingers have the largest radial and ulnar deviation strength as compared to the ring and little fingers. The difference between the first two fingers is not significant. For the female group, the deviation strengths are descending from index, long, right to the little finger. In general, the little finger has stronger ulnar deviation strength than radial deviation strength. These findings are intimately related to the size of the intrinsic muscles. In our evaluation, the force was measured at the PIP joint while all the results of Matheson's study were measured against the finger tip. Therefore, the strength produced by Matheson's study was reduced since the moment arm of the applied forces to the MP joint is larger. In addition, the instability occurring at the DIP and PIP joints must also influence the final strength since in lateral deviation, not only the intrinsic muscles but also the extrinsics must react to maintain the equilibrium. In order to isolate the effect of the distal joints, we chose to measure the deviation strength directly at the PIP joint.

Substantial amounts of data on normal subjects must be collected using the presently developed instruments in order to establish a reliable reference base to fully describe the complex hand function. Such data should also be correlated with sex, age, and physical variations. A large patient series should also be studied so that different pathological conditions can be correlated with associated functional loss. Accurate preoperative assessment and post-operative evaluation can then be performed for objective treatment and care of disabled and handicapped hand patients. Presently, the instruments and methodology are well established. To accomplish the aforementioned objections, well organized study programs with large patient populations are necessary.

REFERENCES

1. Bechtol, C.O., "The use of a dynamometer with adjustable hand spacings", *J. Bone Joint Surg.,* **36A**: 820 (1954).
2. Dickson, R.A. and Nocolle, F.V., "The assessment of hand function. Part I - Measurement of individual digits", *The Hand*, Vol. **4**, No. 3., (1972).
3. Lewey, F. H., Kuhn, W.G. and Juditski, J.T., "A standardized method for assessing the strength of hand and foot muscles", *Surg. Gyn. Obst.,* **85**: 785 (1947).
4. Mannerfelt, L., "Studies on the hand in ulnar nerve paralysis", *Acta Orth. Scand. Suppl.,* **87**: 63(1966).
5. Matheson, A.B., Sinclair, D.C. and Skene, W.G., "The range and power of ulnar and radial deviation of the fingers", *J. Anat.,* **107**: 439–458 (1970).
6. Regnier, J.B., *Consideration sur la force musculaire, suivies de la description et de l'exposition chalcographique d'un nouvel instrument pour mesurer cette force,* (Diss, Paris, 1807).
7. Swanson, A.B., Matev, I.B. and de Groot, G., "The strength of the hand", *Bull. Prosth. Res., Fall,* (1970).
8. Walker, P.S., Davidson, W. and Erkman, M.J., "An apparatus to assess function of the hand", *J. Hand Surg.,* **3-2**: 189–193, March (1978).
9. Schmidt, R.T. and Toews, J.V., "Grip strength as measured by the Jamar dynamometer", *Arch. Phys. Med. & Rehab.,* pp. 321–327, May (1970).
10. Thorngren, K.G. and Werner, C.O., "Normal grip strength", *Acta Orthop. Scand.,* **50**: 255–259 (1979).
11. Burmeister, L.F. and Flatt, A.E., "The prediction of hand strength in elementary school children", *The Hand,* **7-2**: 123–127 (1975).
12. Murray, J.F., Carman, W. and MacKenzie, J.K., "Transmetacarpal amputation of the index finger: A clinical assessment of hand strength and complications", *J. Hand Surg.,* **2-6**: 471–481, Nov. (1977).
13. Armstrong, T.J., Cheffin, D.B. and Foulke, J.A., "A methodology for

documenting hand positions and forces during manual work", *J. Biomech.*, **12**: 131–133 (1979).

14. Smidt, G.L., Flatt, A.E. and Stephens, R.I., "The influence of wrist position on the force produced by the finger flexors", *J. Biomech.*, **8**: 301–306 (1975).

15. Rasch, P.J., Pierson, W.R., O'Connell, E.R. and Hunt, M.B., "A study of the total proportional strength of young adult males", *J. Phys. Med. & Rehab.*, **14**: 369–376 (1960).

16. Everett, P.W. and Sills, F.D., "The relationship of grip strength to stature, somatotype components, and anthropometric measurements of the hand", *The Res. Quart.*, pp. 161–166 (1951).

17. Plato, C.C. and Norris, A.H., "Bone measurements of the second metacarpal and grip strength", *Human Biology,* **52:1**: 131–149, February (1980).

18. Hall, S.J., "Metacarpophalangeal flexion strength with respect to age, sex and exercise habits", *Med. Sci. in Sports & Exercise,* **13–5**: 329–331 (1981).

19. An, K.N., Cooney, W.P., III, Chao, E.Y. and Linscheid, R.L. "Functional strength measurement of normal fingers", *Advances in Bioengineering,* pp. 89–90, ASME (1978).

20. Opitz, J.L. and Linscheid, R.L., "Hand function after metacarpophalangeal joint replacement in rheumatoid arthritis", *Arch. Phys. Med. & Rehab.*, **59**: 160–165 (1978).

6

ABNORMAL HAND JOINT FORCES UNDER SIMULATED PATHOLOGICAL CONDITIONS

A three-dimensional model of the finger, based on accurately defined anatomical structures, has been established and utilized for determination of internal forces under different isometric hand functions. This model was utilized to analyze various simulated abnormal hand conditions, including malunions of phalangeal fractures, tendon and joint injuries, arthrodesis of the DIP joint, and neuromuscular lesions. The methods of simulating these abnormal conditions are described. The predicted internal forces in the hand under such pathological conditions are presented and discussed. The present chapter is only preliminary in nature and is aimed at exploring the feasibility of the model and method of analysis proposed. Therefore, the data presented herein can only be interpreted in a rather limited fashion. Practical clinical application awaits a more complete exclusive investigation.

INTRODUCTION

Biomechanical study of the finger during various hand functions in normal and pathological states provides a sound basis for the development of finger joint prostheses and the rationalization of reconstructive procedures in hand surgery. Flatt and his associates[1,2] performed a series of biomechanical analyses of finger functions pertaining to the pathomechanics of ulnar drift in rheumatoid arthritis and the rational development of the metal hinge prostheses. A mathematical finger model was constructed by using tri-articular linkage with monaxial joints. The articulating surfaces were assumed to be circular in two dimensions and the functions investigated were limited to free flexion and extension. The effects of dorsal and volar drift of the MP joint axis on the digit motion and tendon force balance were analyzed. However, these studies were primarily planar analyses, which were inadequate in that they ignored the lateral constraining forces of ligaments in static functions and joint mobility. The rotational and lateral deviation in kinematic analyses was also overlooked. With such planar analyses, it would be difficult to adjust for the pathological changes occurring in realistic clinical situations. Furthermore, only three-dimensional analyses contain sufficient flexibility to allow the system to accommodate anatomical changes that may be introduced by intricate reconstructive procedures.

In the present study, a static three-dimensional mathematical model of the hand based on the forces and moments provided by both intrinsic and extrinsic muscles and their tendinous elements acting across the finger joint was developed. This allows theoretical determination of tendon forces, joint constraint forces, joint moments, and joint contact forces based on the applied functional loads acting on the fingers. These internal forces and moments can then be related to the various output forces measured in normal subjects. This mathematical model of the hand in static equilibrium was modified to adopt various abnormal states of the hand, including malunions of fractures, tendon subluxation, and neuromuscular weakness and paralysis. Biplanar roentgenography for joint orientation and tendon location analysis and direct force output measurements in patients were applied in the present investigation. The pathological changes in the anatomical structures were estimated based on clinical observations.

METHODS AND MATERIALS

In this study, four abnormal conditions of the hand were investigated. Since these simulated pathological conditions did not consider soft tissue location changes in relation to the joint, the mathematical analysis based on the normative hand model developed previously was used. In severe rheumatoid arthritis and other neuromuscular lesions involving significant changes in soft tissue structure and joint orientation, different models incorporating these structural alterations must be used. Simulated abnormal cases are briefly described as follows.

The first case considered the reduction of muscle strength. Such abnormality can be caused by traumatic injury or neuromuscular diseases. All extrinsic and intrinsic muscles were studied. Systematic reduction of muscle strength, by assigning appropriate values for the upper bounds of muscles, was introduced, and the corresponding functional pinch and grasp forces were calculated. The best selection of donor muscles to restore function with low ulnar nerve palsy was examined based on the PCSA data in Chap. 2. In the case with complete laceration of the FP, DIP joint fusion was considered. The effects of joint fusion on finger force were studied under a separate category.

The second case concerns tendon subluxation. Such a lesion occurs when the constraining pulley system is damaged by trauma or rheumatic disease. This abnormality was simulated by increasing the distance between the tendon and the joint center of rotation in the normal model. The magnitude of the distance change was altered gradually to observe its effect on finger functional strength. Mathematically, such change can easily be incorporated by altering the coefficients in the governing equilibrium equations. FP and FS were studied separately and together for volar subluxation only. The long extensor and radial interosseous muscle were examined for radial subluxation.

The third category analyzes the effect of malunion of phalangeal fracture. Such lesion may produce angular deformity of the phalanx and bone shortening. In either case, reduction of muscle strength and hand function would result. Only the proximal phalanx was considered at the present time. The specific abnormality was simulated by introducing angulation along the length of the long bone. Tendon location and joint orientation at its distal and proximal ends were assumed unchanged.

In the final case, the effects of DIP joint fusion angle on finger functional strength were studied. Such a procedure is required following laceration of the profundus tendon. Different fusion angles were tried and their associated tip pinch forces were calculated. The applied pinch force was also varied so that its effect on the final function of the finger could be analyzed.

The mathematical model used to simulate these abnormalities was the same as that used for the determination of normal muscle and joint forces in Chap. 3. However, in order to calculate the maximal applied force or strength of a specific function, the magnitude of the externally applied force was treated as an additional unknown variable as well, and used as the objective function for optimization procedures.

RESULTS

The results for the abnormal conditions analyzed, based on simulated models of the index finger, are presented separately. When no joint deformation was involved, the same orientation angles as used in the normative model were utilized.

A. *Analysis of Muscle Strength Reductions and Tendon Transfers*

This case provided simulated analysis of neuromuscular weakness and partial or complete rupture of the tendons. The resulting tip pinch force reductions due to various extrinsic and intrinsic muscle strength losses are illustrated in Figs. 6–1 and 6–2. With the FP completely inactive, no pinch force could be generated, since the DIP joint was no longer able to maintain its joint configuration. A similar effect was also found when the UI had no strength. The inability to maintain equilibrium at the MP joint in the radio-ulnar direction might have been the main reason. Substantial strength loss was noticed if these

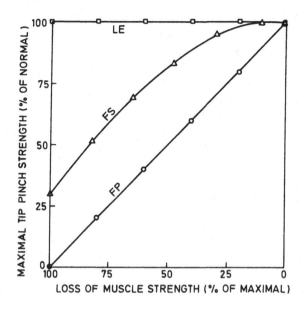

Fig. 6-1 Reduction of tip pinch force due to various extrinsic muscle strength loss.

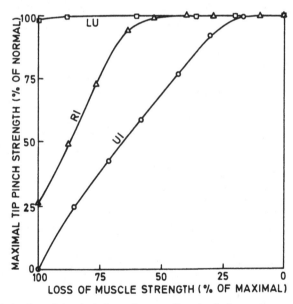

Fig. 6-2 Reduction of tip pinch force due to various intrinsic muscle strength loss.

muscles were partially active. The pinch and grasp functional force reductions were found more sensitive to extrinsics than to intrinsics.

When the FS tendon was completely lacerated, the index finger could still perform tip pinch with the compensation of higher contractive forces of other muscles (Table 6–1). These tendon and muscle forces are in units of applied tip

Table 6-1 Muscle forces in index finger for tip pinch with and without laceration of flexor sublimis (FS).

Tendon/Muscle Force
(unit of applied pinch force)

	Normal	FS Laceration
FP	2.20	5.46
FS	1.77	- -
RI	1.26	1.43
LU	0.15	1.85
UI	0.74	2.37
LE	- -	7.38

pinch force. Substantial increases of normalized muscle forces indicate that the estimated pinch strength will be significantly reduced. This compensatory mechanism among the remaining available muscles to maintain the maximum possible functional strength when one of the muscles is partially, or completely, inactive is further illustrated in Fig. 6–3. In this case, the flexor profundus and long extensor become more active, although other intrinsic muscles also play a part in compensation by increasing their contractive forces.

The loss of tip pinch strength when one, or more, muscles have lost their function completely is summarized in Fig. 6–4. The shaded area reflects the amount of tip pinch strength loss. These results were obtained by assuming the MP joint as a universal joint where mediolateral equilibrium was required. When an ulnar nerve lesion produces paralysis of both the first dorsal and the radial interosseous muscles, tip pinch would be reduced to about one-sixth of its normal value. Inability to perform normal tip pinch function should be a typical observation for this type of lesion. The selection of various donor muscles for tendon transfer to the RI to restore tip pinch strength is illustrated in Fig. 6–5. In radial nerve lesions with total loss of extensor muscle strength, tip pinch action was not affected appreciably. Under high median nerve injury, both FP and FS are paralyzed; tip pinch under normal hand con-

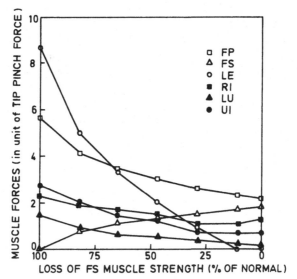

Fig. 6-3   Compensatory mechanism among the remaining available muscles for maximal tip pinch force when FS muscle is affected.

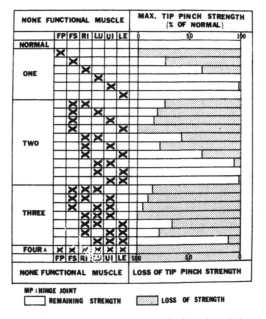

Fig. 6-4   Tip pinch strength lost when one or more muscles have lost their function completely.

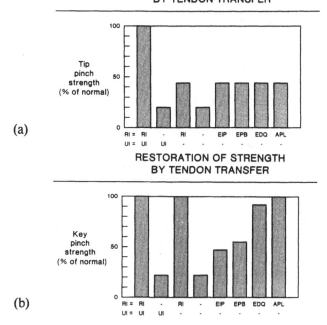

Fig. 6-5 (a) Tip pinch strength attainable with tendon transfer to RI.
(b) Key pinch strength attainable with tendon transfer to RI.

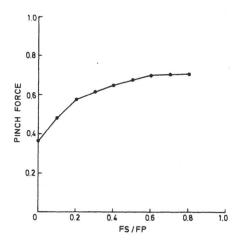

Fig. 6-6 The possibility of recovering the tip pinch force by splitting the FS muscle to attach to the paralyzed FP tendon.

figuration was not possible. However, if the median nerve lesion occurs at a more distal level (AIO) with FS function intact, certain tip pinch strength can be restored through muscle transfer. The possibility of recovering pinch force by splitting the FS muscle to attach to the paralyzed FP tendon is illustrated in Fig. 6-6. Seventy percent of normal tip pinch strength can be restored by distributing one-fifth of the FS muscle to the FP tendon.

In addition to the theoretical analysis, experiments have also been carried out to stimulate tendon transfer for opposition of the thumb. The effective moment arms for abduction and flexion of the first metacarpal were calculated *in vitro* by biplanar X-ray technique to determine the effect of tendon transfer on thumb motion and strength (Fig. 6-7a). From these studies, the contribution of thumb muscles and joints to opposition of the thumb to the fingers was determined. The abductor pollicis brevis was found to be the most important thumb muscle providing this action. The opponens pollicis was an important contributory muscle. Optimum prehensile strength and motion was obtained through transfer of flexor tendons taken from the ring or long finger, or transfer of an extensor tendon (ECU, ECRL) taken from the wrist, passed around, or near, the pisiform bone, and directed into the radial side of the proximal phalanx of the thumb (Fig. 6-7b).

B. Analysis of Tendon Subluxation

Tendon subluxation may be produced by trauma, rheumatoid disease, and even by surgical procedure, such as pulley advancement for the correction of claw finger deformity. In these circumstances, tendons will not remain in their normal position with respect to the joint geometry, thereby causing force changes in the tendons and muscles, as well as the joint. Volar subluxations of the FS and FP in the index finger were analyzed, and the associated tendon and joint forces are tabulated in Tables 6-2 and 6-3. The amount of subluxation was expressed in terms of the middle phalanx length. Up to a subluxation of 6% of the mid-phalanx length was analyzed. Significant reductions in muscle and joint forces required for the same tip pinch strength were noted.

Therefore, the maximum tip pinch strength in the index finger was expected to increase with volar subluxation of both flexors at the MP joint (Fig. 6-8). However, such an insignificant increase of functional strength may be nullified by the expected joint instability associated with tendon subluxation. Therefore, the tip pinch strength increase observed in this case should not be overemphasized.

Both the long extensor and the first interosseous are important contributors in lateral pinch. Radial subluxation of these tendons at the MP joint would certainly benefit this function. Figure 6-9 demonstrates this effect by showing the proportional reduction in these muscles with radial subluxation of the

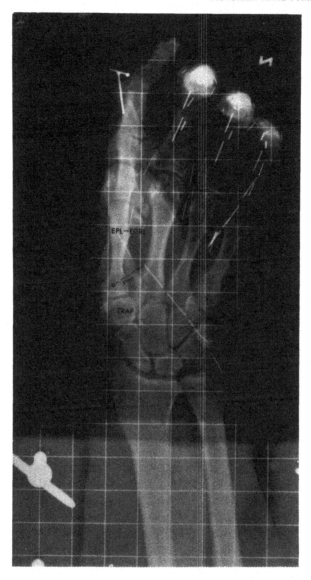

Fig. 6-7 (a) Biplanar X-ray, utilizing a special X-ray grid technique.

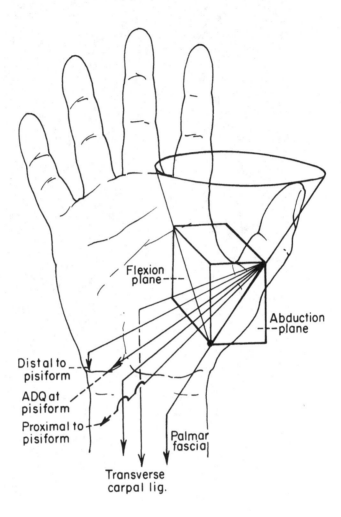

Fig. 6—7 (b)   Thumb opposition transfer requires moment arms in abduction and flexion planes
directionally toward the pisiform. Effectiveness of transfers, such as ECRL, can
be calculated from biplanar X-rays.

Table 6-2 Forces of muscles in index finger for tip pinch with subluxation of FP and FS at MP joint at various levels. The force is expressed in units of applied force.

	Tendon/Muscle Force			
Volar Subluxation of FP and FS (in units of middle phalanx length)	0.00	0.02	0.04	0.06
FP	2.197	2.156	2.116	2.076
FS	1.774	1.796	1.819	1.842
RI	1.256	1.277	1.302	1.330
LU	0.147	0.103	0.059	0.016
UI	0.736	0.670	0.603	0.538
LE	- -	- -	- -	- -

Table 6-3 Joint constraint force in index finger for tip pinch with subluxation of FP and FS at MP joint at various levels. The force is expressed in units of applied force.

		Joint Constraint Force			
Volar Subluxation of FP and FS (in units of middle phalanx length)		0.00	0.02	0.04	0.06
DIP	F_x	2.96	2.87	2.78	2.70
	F_y	0.40	0.38	0.35	0.33
	F_z	$-.14$	$-.14$	$-.14$	$-.13$
PIP	F_x	4.92	4.82	4.73	4.64
	F_y	1.09	1.06	1.04	1.01
	F_z	$-.03$	$-.03$	$-.02$	$-.02$
MP	F_x	4.09	3.98	3.86	3.75
	F_y	3.04	3.00	2.97	2.95
	F_z	0.47	0.50	0.54	0.57

extensor tendon under identical pinch strength. The variation of these forces was due to the different objective functions used in solving the indeterminate problems of force analysis. Again, the amount of subluxation is expressed in units of the middle phalanx length. A significant amount of muscle strength reduction (LE by 14%, RI by 13%) is possible with a subluxation of the extensor radially at a distance of 5% of the middle phalanx length.

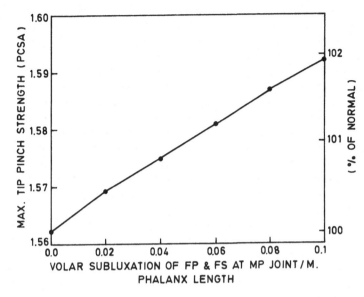

Fig. 6-8   Maximal tip pinch force in the index finger with various amounts of volar subluxation of both flexors at the MP joint.

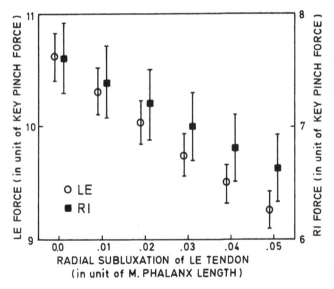

Fig. 6-9   Reduction of the forces in RI and LE muscles with radial subluxation of the LE tendon under identical pinch strength.

C. *Analysis of Phalangeal Fracture Malunion*

As an example, the present analysis only considered the malunion of the proximal phalanx of the index finger. Such deformity causes rotation of the coordinate axes fixed to the distal and proximal portions of this long bone. However, tendons at both the PIP and MP joints were assumed to maintain their normal orientation with respect to the joints. The malunion angle Φ was found to affect tip pinch force significantly, as demonstrated in Fig. 6-10. Apparently, such variation of pinch force also depends on the pinch force direction. No consideration was included for bone shortening, and thus the tendon excursion may be a result of such deformity. This effect may be significant and should be included in future investigations.

The results show that pinch force would increase regardless of the sign of angle Φ (concave or convex angulation) if the applied pinch force is oriented at an angle of less than 90° with respect to the long axis of the distal phalanx. Such force increase is basically caused by the reduction of the moment arm of the externally applied force with respect to the joint center of rotation when the proximal phalanx is angulated. The precise clinical implication of such an abnormality is difficult to correlate at this moment, but it indicates that

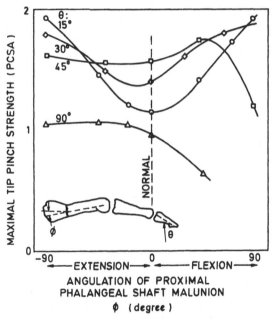

Fig. 6-10 The effect of proximal phalanx fracture malunion on the tip pinch force of the index finger.

malunion of the proximal phalanx at a small angulation has little effect on the functional strength of pinch. Additional anatomical and clinical studies must be performed in order to substantiate this result.

D. Analysis of Optimal Fusion Angle at the DIP Joint

When the FP is lacerated but not repaired nor transferred, the DIP joint becomes unstable and no pinch action can be produced in the normal configuration. DIP joint fusion was considered as one of the possible surgical procedures for this condition. The required tendon and joint forces during tip pinch after such a procedure were simulated and summarized (Table 6-4). The graph shown in Fig. 6-11 seems to suggest that a smaller fusion angle is more

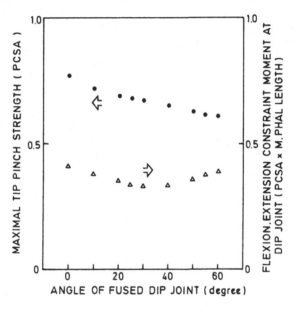

Fig. 6-11 The effects of fusion angle at the DIP joint on the tip pinch strength and the bending moment at the joint.

desirable to maintain pinch strength. However, if the stability of a fused joint is considered, a 30° angle would be more appropriate to minimize the bending moment applied at the DIP joint which may cause pin deformity and nonunion.

Table 6-4 Muscle force in index finger for tip pinch with laceration of FP and DIP joint fused at various angles. The force is expressed in units of applied force.

Fusion Angles of DIP Joint	Tendon/Muscle Force					
	0°	10°	20°	30°	40°	50°
FP	—	—	—	—	—	—
FS	3.63	3.92	4.11	4.21	4.31	4.52
RI	1.55	1.47	1.34	1.17	1.07	1.18
LU	—	—	—	—	—	—
UI	.79	.61	.41	.17	—	—
LE	—	—	—	—	.40	1.76

DISCUSSION

The abnormal models and the results presented in this chapter were aimed at exploring the feasibility of using mathematical simulation to study the patho-mechanics of the hand under a variety of conditions. Although four representative cases were investigated, other types of lesions can also be analyzed if their anatomical and functional deficiencies could be defined quantitatively. These cases approximated the actual pathological changes involved with primary emphasis on the biomechanical aspects of the problem. Not all of the anatomical involvements or important clinical considerations were incorporated into the present model. Therefore, the application of the current results to clinical consideration should be judged carefully with respect to the assumptions made in the theoretical models. Nevertheless, the knowledge obtained from the present analysis should help orthopedic surgeons and physical therapists to properly evaluate not only patients but also treatment methods. It is also possible that an extension of such a study could provide guidelines for future improvement of surgical techniques, implant designs and rehabilitative therapy so that the functional ability of disabled and handicapped hand patients could be restored optimally.

Although the precise anatomy involved in each type of pathology cannot be duplicated in the analytical model, the correlation between the amount of deformity and the internal force changes can establish a valid pattern to study the biomechanical effects of such lesions. The present analysis allows a systematic variation of the extent of lesion and proceeds to estimate the effect such a lesion has on the functional performance of the hand, which is impossible to study anatomically. This is one of the main advantages of investigating the pathomechanics of the hand based on mathematically simulated models. Such a procedure can be carried one step further to evaluate the effects of treatment based on objective criteria. Without actually performing the procedure, its outcome on restoring hand function can be predicted on a comparative basis. The value of the present analysis lies precisely in this respect.

Due to the large variation of pathological conditions involved in the hand, the most common and well-defined abnormalities should be analyzed first. Complete assessment of clinical and pathlogical changes associated with each abnormality must be quantitated by roentgenographical, neurological and functional evaluations in a large patient series. A careful study of currently available surgical and therapeutic treatments in the hand is essential in order to rationalize the clinical concepts and to properly guide the future investigations.

REFERENCES

1. Flatt, A.E. and Fischer, G.W., "Biomechanical factors in the replacement of rheumatoid finger joints", *Ann. Rheum. Dis. Suppl.* **28**: 36–41 (1969).
2. Flatt, A.E., "The pathomechanics of ulnar drift", *Social and Rehabilitation Services Final Report*", Grant No. RD 2226 M (1971).

7

QUANTITATIVE EMG ANALYSIS OF HAND EXTRINSIC AND INTRINSIC MUSCLES DURING ISOMETRIC FUNCTION

INTRODUCTION

The active forces during isometric functions of pinch and grasp cannot be directly measured at the present time. Musculotendinous and joint forces, which can be determined from mathematical analysis of modeled mechanical finger and thumb joint systems, are difficult to verify without an experimental method that can directly measure force output *in vivo*. Electromyographic (EMG) activity is a readily available technique which may have quantitative significance in studying the amount of force exerted in muscles under isometric contractions. This study used quantitative EMG to substantiate and further analyze hand functions as predicted by the mathematical model.

Many investigators have explored the correlation between EMG and muscular force since 1947[3,13]. There is definite agreement that a linear relationship exists between EMG and tension within a specific muscle provided that the muscular effort is isometric, in the mid-range of exertion, and that muscle fiber length remains unchanged. Further restrictions are that there is a standard speed of muscle contraction, the electrode positions remain constant, and that muscle fatigue is minimized. The quantitative results are best expressed as percents of maximum integrated EMG activity per muscle studied. Inman *et al.*[7] first demonstrated the ability to compare isometric tension and EMG through the use of a strain gauge dynamometer attached to a biceps cineplasty. The electrical potentials were rectified and integrated to produce an average EMG signal. He showed that electrical output closely parallels the tension in a given contracting muscle, provided that the contraction is isometric. The concept of rectification and integration of the complex EMG signal was an important step forward, which enabled the study of the relationship between muscle tension and the summated EMG wave signals since the interference pattern produced by direct or "raw" EMG recordings prevented accurate quantitative analysis.

Lippold[9] reported a direct linear relationship between surface electrode activity and a foot-plate dynamometer. Bigland and Lippold[1] extended this concept to isotonic contractions in an effort to analyze the muscle force velocity-time relationship with EMG. Hagberg and Johnson[5] showed a linear relationship between the time integral of the full-wave rectified myoelectric signal amplitude of the biceps muscles and performed work consisting of intermittent static elbow contractions. However, other attempts to correlate both isometric and muscle functions with EMG activity under isometric-anisotonic and isotonic conditions having shown a nonlinear relationship. Komi and Buskirk [22], Metral and Cassar[23], and Zuniga and Simons[21] relate this to the failure to record EMG-tension relationship through the entire range of muscle contraction. This latter group employed an x-y recorder to measure averaged EMG potentials against sustained isometric tension. They found a nonlinear relationship fitting the curve by a quadratic regression line but not by a linear regression line. In the midrange of testing with sustained contraction, however, their curves were linear and the coefficients of variation were quite small. Perry[12,20] pointed out the difficulties in measuring muscle force by EMG with both isotonic and isometric testing. Maximum muscle force contraction, or any change in muscle resting length, negated the direct correlation of integrated EMG and muscle tension. For a given force, it was shown that the EMG was larger when the muscle was tested in a shortened position and relatively small if the fibres were stretched[7].

Most recent studies support the contention that increased muscle tension corresponds to the increased number of individual motor units and to the frequency of the action potentials of the involved muscle[16-19, 22]. They state that such relationship can be linear. Surface electrodes and intramuscular electrodes, either individually recorded or algebraically summed, were able to obtain constant values of EMG within the same muscle. Double observations and repetitive testing of the same muscle also showed that reproducibility existed within individual subjects, but that results between individuals might not be comparable. Wire electrodes compared favorably with surface and needle electrodes, and for small muscles wire electrodes appeared to be superior. Their sampling area was small, however, and in the versatile and complex muscles of the hand, more testing is required to demonstrate consistent recordings.

Two major difficulties encountered in the use of EMG to quantitate muscle tension were recognized by previous investigators[27]. First, in order to calculate the force, the moment arm through which the force acts must be determined. Second, the contribution of both agonist and antagonist muscle forces should be included. Simultaneous determination of these muscle forces was developed for force distributions in the elbow[29,30,31]. A similar procedure has thus been developed and used in this study.

METHODS AND MATERIALS

Two types of electrodes were used to provide comparative data and to accurately record the size variation present between extrinsic and intrinsic muscles. Surface electrodes consisting of copper discs, 0.5 cm in diameter, were applied with a gel paste. They were securely taped over each test muscle and separated by 2 to 6 cm intervals, depending on the muscle size. The Karma wire electrodes[24,25,26] were made of 36 gauge enameled copper with bared tips, hooked over a distance of 5 mm (Fig. 7-). They were inserted through a 23 gauge needle which was withdrawn after satisfactory verification of the proper electrode placement was established. Electrode placement was achieved by using the cross-sectional anatomy described in Chap. 2 as a guide and was confirmed by isolated muscle function testing and electrical stimulation. The slight time lag from signal output through the filtering stage to the electrical recording on paper or screen were not significant for isometric testing and can be disregarded. To further reduce noise, a two-pass processor[12,20] was utilized whereby the averaged signal was subtracted from each sample to correct for any low frequency drift or DC bias in the waveform. Rectification of the conditioned EMG signal and subsequent integration and averaging were performed by a specially designed analog device (circuit diagram shown in Fig. 7-2). Simultaneous functional strength and EMG measurements of intrinsic

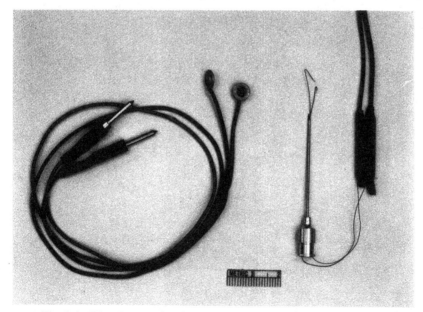

Fig. 7-1 Wire electrodes and the implantation needle used by Basmajian.

and extrinsic muscles were obtained during each experimental test (Fig. 7–3) and were recorded on a digital computer via an analog to digital converter.

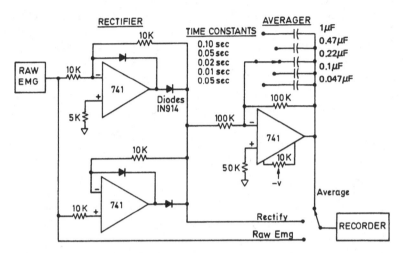

Fig. 7–2   Electronic circuitry for EMG rectification and averaging (RMS value) processes.

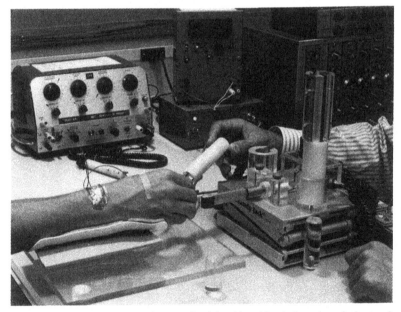

Fig. 7–3   EMG and functional force evaluation of hand intrinsic and extrinsic muscles.

Instrumentation utilized in this study included a 12–channel Honeywell Visicorder (Fig. 7–3), a Tektronix oscilloscope, a Hewlett-Packard *X-Y* recorder (Fig. 7–4), and a computer terminal to communicate with the PDP 11/34 computer. Eight muscles could be evaluated simultaneously. The forces were measured with a set of pinch, grasp and lateral deviation force transducers.

Fig. 7–4 Hand muscle EMG testing and recording instruments plus the functional strength measurement device. (A) Hewlett-Packard X-Y recorder, (B) Tektronix oscilloscope, (C) EMG amplifier, (D) analog integrator, (E) computer terminal, (F) force readout device.

TESTING PROCEDURE

After proper placement of the electrodes and optimal selection of the amplification gain for individual channels was achieved, the experiments were carried out in two parts.

A. Isolated Isometric Test

The purposes of this procedure were first to derive the maximum EMG signals of individually tested muscles for future data normalization and second to establish the muscle tension and EMG relationship. The tests were performed with all of the thumb or finger joints in the neutrally extended position. The isometric flexion, extension, abduction and adduction strengths were measured sequentially at the levels of the DIP and PIP joints. The forces applied on the transducers were displayed on an oscilloscope as visual feedback for the test subjects to control their contraction speeds. The EMG signals of each muscle were simultaneously recorded by computer and visicorder.

B. Pinch and Grasp Function Test

In this part of the study, the forces between the thumb and index finger were measured simultaneously with EMG signals during pinch and grasp functions. With the data from this experiment, individual muscle EMG activities, and thus the tension, could be correlated with that predicted from the theoretical models based on the identical pinch and grasp forces.

RESULTS

The thumbs of seven normal subjects were tested. From these studies, bipolar wire electrodes were found to be superior to surface electrodes in

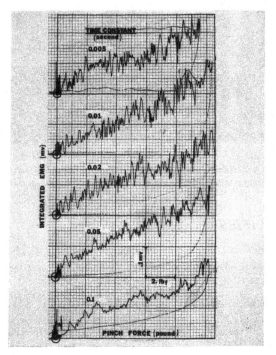

Fig. 7-5 Integrated EMG of the APB muscle of the thumb during pinch based on different time constants for signal averaging and integration.

recording the EMG activity in the small muscles of the hand. Repeated tests of subjects on different days demonstrated close reproducibility of the force to EMG relationship, provided electrode placement was nearly identical. Electrodes retested up to 30 minutes after an initial force test produced identical recordings and suggested that repeated testing does not produce muscle fatigue. Times constants for the signal averaging and integration tested over a range of 0.005 to 0.1 seconds demonstrated the best choice to be 0.05 seconds, which provides the smoothest signals with negligible distortion (Fig. 7-5).

Typical recordings of rectified-integrated EMG signals of the eight thumb muscles versus the applied forces are illustrated in Figs. 7-6 through 7-11 for

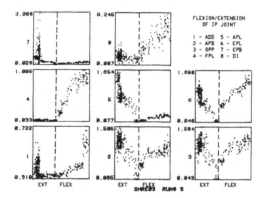

Fig. 7-6 The integrated EMG (vertical axis) versus the applied flexion/extension force (horizontal axis) on the distal phalanx of the thumb. The unit used for EMG magnitude is in MV.

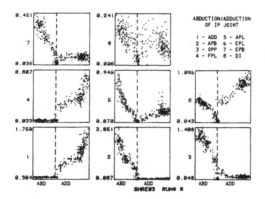

Fig. 7-7 The integrated EMG (vertical axis) versus the applied abduction/adduction force (horizontal axis) on the distal phalanx of the thumb.

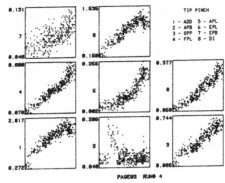

Fig. 7-8 The integrated EMG (vertical axis) versus tip pinch force (horizontal axis) in tip pinch function.

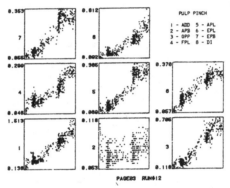

Fig. 7-9 The integrated EMG (vertical axis) versus pulp pinch force (horizontal axis) in pulp pinch function.

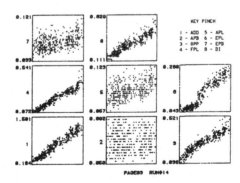

Fig. 7-10 The integrated EMG (vertical axis) versus key pinch force (horizontal axis) in key pinch function.

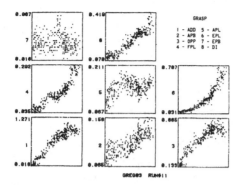

Fig. 7-11 The integrated EMG (vertical axis) versus grasp force (horizontal axis) in grasp function.

flexion-extension, abduction-adduction, tip pinch, key pinch, pulp pinch and grasp, individually. In each graph, the horizontal axis represents the force and the vertical axis plots the EMG activity. From these graphs, at any given functional force level, the relative EMG activity or "tension" of individual muscles can be evaluated in terms of the percentage of maximal effort.

To determine the relationship of muscle tension and its integrated EMG, an algorithm based on the theoretical force analysis model discussed in Chap. 3 was developed (Fig. 7-12).

MUSCLE FORCE PREDICTION FROM EMG

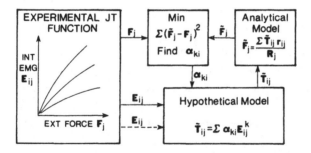

Fig. 7-12 Mathematical model for the determination of the muscle EMG and tension relationship. $\tilde{T}_{ij}$ is the estimated muscle tension and $\tilde{F}_j$ is the estimated applied force based on the moment arm for the muscle (r_{ij}) and for the applied load (R_j).

In this procedure, the estimated ith muscle tension under jth function ($\tilde{T}_{ij}$) and its integrated EMG (E_{ij}) were assumed to be related by a polynomial function,

$$\tilde{T}_{ij} = \sum_{k=0}^{n} \alpha_{ki} (E_{ij})^k \tag{1}$$

where α_{ki} is the coefficient for the kth order term of the polynomial in terms of E_{ij}. If these relationships were substituted in the moment equation, then the estimated muscle tension can be related only to the integrated EMG signal and the muscle and load moment arms based on a nonlinear relationship in terms of the coefficient α_{ki}. Through multiple polynomial regression analysis (least-squared minimization process), the coefficient, α_{ki}, in equation (1) for the ith muscle was calculated with the experimental data obtained during isolated isometric function testing.

Multiple polynomial regression up to the third order was performed. To determine the closeness of regression fit, correlation coefficients were used. In general, the first order linear relationship provided a reasonable fit ($R = 0.86 \sim 0.99$). Adding more terms did not improve the results significantly. The coefficients based on the linear relationship for the 8 thumb muscles in 5 test subjects are summarized in Table 7-1. The variation among different tests in an individual subject is small, but the variation between different subjects is high. The predicted muscle tension can be estimated based on the scale factor of 89 N/volt.

Table 7-1   The coefficients of the linear relationship between muscle tension (volt) and processed EMG (volt).

MUSCLE	REM1	SHRE	SUBJECT TONI	PAGE	GREG
ADD	.42 (.13)	.09 (.02)	.49 (.10)	.24 (.09)	.32 (.02)
APB	.24 (.07)	.05 (.01)	.12 (.02)	.16 (.07)	.29 (.03)
OPP	.06 (.04)	.04 (.01)	.06 (.02)	.09 (.04)	.06 (.02)
FPL	1.23 (.20)	.62 (.03)	.54 (.10)	.62 (.03)	2.01 (.28)
APL	.29 (.06)	.05 (.03)	.08 (.06)	.06 (.01)	.07 (.01)
EPL	.48 (.04)	.19 (.03)	.18 (.05)	.53(.08)	.21 (.01)
EPB	.11 (.01)	.07 (.02)	.06 (.04)	.17 (.05)	.16 (.07)
DI	.30 (.25)	- -	.08 (.04)	.29 (.16)	.20 (.09)

Mean (S.D.) of four runs.

Absolute forces of the eight muscles during different pinch and grasp functions were calculated by using the recorded integrated EMG combined with the polynomial relationships (Table 7-2). The results are normalized against the applied force. They were calculated by averaging the muscle force/strength ratio at the middle-half of strength range. Large variations of the muscle forces were observed. On the average, muscle forces ranged from 4 N to 40 N for the EPL and 3 N to 120 N for the FPL muscles in different isometric functions. These are within the range predicted in the theoretical model previously reported[32].

Table 7-2 Thumb muscle forces during pinch and grasp* (in units of applied force).

MUSCLE	TIP PINCH	PULP PINCH	KEY PINCH	GRASP
ADD	0.38 – 15.4	0.6 – 8.4	1.4 – 4.5	0.5 – 11.0
APB	0.0 – 1.45	0.0 – 0.7	0.0 – 1.6	0.0 – 0.8
OPP	0.6 – 2.6	0.3 – 2.3	0.5 – 3.3	0.4 – 3.1
FPL	1.1 – 3.5	1.7 – 3.8	0.9 – 1.7	0.2 – 1.7
APL	0.0 – 0.4	0.2 – 0.5	0.0 – 0.6	0.0 – 0.5
EPL	0.4 – 3.0	0.3 – 1.6	0.5 – 2.3	0.2 – 0.8
EPB	0.0 – 0.5	0.0 – 0.3	0.2 – 0.7	0.0 – 0.1
DI	0.2 – 8.1	0.2 – 2.4	0.3 – 2.1	0.2 – 4.1

* Based on the EMG study of five subjects

DISCUSSION

Electromyography has been used extensively to study active muscle function in the motion of finger and thumb joints but rarely has been applied to measure muscle tension. Forrest and Basmajian[4] tested the thenar and hypothenar intrinsic muscles but only in isotonic actions of opposition, extension and flexion. Their static testing was limited to clasping a wooden dowel and holding a cup with no accurate means of measuring the strength involved in each muscle responsible for the activity. Integrated EMG signals were not utilized in assessing their results.

Close and Kidd[2] studied synchronous recordings of thumb, index and long finger motions and the muscle action potentials. This included static grasp but not pinch; separate evaluation of individual muscles was, therefore, impossible. Furthermore, their work concerned mainly dynamic hand function.

Hall and Long[6] analyzed static pinch and power grip in a well-planned study but limited the muscles tested to only the intrinsics. Various grip and grasp actions were tried, but the magnitudes of these activities were not accurately measured. In addition, no clear distinctions were made between power grip and precision handling, as emphasized by Landsmeer[8] and Napier[15]. No attempt was made to quantitate the muscle action potentials.

A series of studies by Long and Brown[10] described the electromyographic kinesiology of long finger muscles. Both intrinsic and extrinsic muscles were analyzed. These papers, however, dealt only with motion of the long finger with no correlation to functional strength and with no isometric strength testing. Similar studies of the index finger were conducted by Boivin *et al.* [14], but the results were restricted to kinematic analysis.

Gydikov and associates[28] determined that the opponens pollicis has three types of motor units which respond to an electrical impulse; a) fast contracting and resistant to fatigue, b) fast contracting and fatigable, c) slow contracting and resistant to fatigue. They explained the tonic and phasic natures of muscle fiber action by determining whether the motor unit was fast acting but fatigable (phased in only at increased muscle effort) or fast and slow acting but unfatigable (tonic) and acting only at lower, but constant, muscle efforts. They did not determine how the tonic and phasic motor types affect resultant forces, or whether the action potentials varied. Their data, however, may explain the sudden rise in EMG electrical activity with summation of motor units at the high load values, and why nonlinear regression curves are seen with faster pinch and grasp rates.

In the majority of these provious studies of muscle function, the EMG action potentials were subjectively measured on a scale system. The works of Inman *et al.*[7] and Bigland and Lippold[1] forcus directly on the problems of quantitative EMG and the specific importance of integration of the EMG signal. These and other authorities have emphasized the requirements of isometric testing which include a) low, midrange and high effort recordings, b) uniform muscle lengths, c) fixed joint positions, d) constant rates and e) subject training. Changing muscle fiber lengths and electrode sampling areas are potential sources of error that can only be prevented by rigid testing procedures so that these parameters remain absolutely constant.

The concept of using polynomial regression analysis for identifying the muscle EMG and tension relationship is a simple technique and provides a better approach for establishing these relationships. As described, joint function results from the contributions of multiple muscles. During isolated joint contraction, proper distribution of individual muscle tension working against the external load is significant in the establishment of a reliable EMG-muscle force relationship. The stability and accuracy of the multiple regression analysis depends on the ranges of both dependent and independent variables. In this particular study, for example, both EPL and FPL are proportionally active in flexion of the IP joint. If only the data from flexion of the IP joint are used for the regression analysis, the coefficient for the EPL in the regression formula (equation (1)) will be the same as that of the FPL. This implies that both the EPL and FPL are functioning as flexors and this is not correct. However, if the data from the extension testing of the IP joint, in which the FPL is not involved, is also included, the coefficients predicted for the regression formula will be consistent and more reliable. Based on the same argument, both abduction and adduction tests are used together in the regression analysis to obtain the coefficients of the abductors and adductors. Simultaneously combining the data from all 4 functions into the regression analysis was not attempted because the mixture of moment arms in two orthogonal planes makes it difficult to calculate the coefficients.

From a physiological consideration, if there is no EMG activity, there will be no active tension created by the muscle. Therefore, in the model for the EMG and tension relationship of muscles, the constant term (i.e. zeroth order term) in the regression polynomial is not included. In fact, it has been examined and the results of calculated regression coefficients and correlation coefficients did not change too much based on the model with, or without, the constant terms. The correlation coefficients in this analysis have been very encouraging. For all of the five subjects studied, the coefficients were between 0.86 and 0.99, even when including only the first-order terms in the model. The coefficients improved (always up to 0.99) when extended from first to second or third order. However, in most cases inclusion of the higher terms did not significantly improve the correlation statistically. This implies that for isometric function at the physiological midrange of strength, the linear relationship between EMG and tension of the muscles is a reasonable model for this type of study.

It should be remembered that the EMG and tension relationships for the models predicted by the regression analysis in this study are more reliable for both FPL and EPL muscles since these are the only two muscles contributing to the flexion and extension motion at the IP joint. The relationship for the other muscles may not be as accurate, since function at other joints involves multiple muscles and the moment arms for these muscles are difficult to measure in each test subject. If the muscle moment arm can be accurately measured *in vivo* with changes in joint position, then the reliability of the relationship is expected to improve.

In addition, the regression coefficients for the EMG and tension model predicted are not universal. These coefficients are valid only for that particular subject during that specific study. The coefficients depend on the EMG signals, which are related to the location and type of electrodes and the gain of the instrument being utilized. However, these coefficients did serve as a reliable index for muscle force calculations in pinch and grasp in the later part of the study, since electrode placement was not changed from single muscle testing to multiple muscle testing.

Large variations in muscular activities are noticed among subjects during various isometric thumb functions. In addition, muscle responses to increments of external load within each individual are also different. Three types of response are normally observed. Group I muscles respond accordingly with the increase of load; these muscles are usually the dominant, or major, primary muscles for that specific function. Group II muscles may not be involved at the lower ranges of external load, but become secondarily active as soon as the strength reaches a certain range. These muscles may act as the reinforcing units to balance the applied force, or they may serve as joint stabilizers once the joint constraints reach beyond ligament or capsule strength. Group III muscles are usually antagonistic muscles which respond only to balance the applied load and maintain joint equilibriums. These studies indicate that as muscles reach their strength limit, further strength or function responsibilities can be made up for by other muscles which actively function until they also reach their own ultimate strength capabilities.

Despite the aforementioned potential difficulties and limitations, it appears that an accurate method of recording muscle action potentials by integrated EMG and simultaneous measurement of external static hand forces can provide qualitative verification of analytically determined force distributions in muscles. In this study, the ranges of forces found in the FPL and EPL muscles (0.1 to 0.3 and 0.1 to 2.3 times the externally applied forces, respectively) were within the range of those predicted by the theoretical calculation. However, additional research is deemed necessary in order to fully understand the relationship between muscle EMG and its contractive force. Such relationship, if properly established, can even provide plausible application in clinical assessment of muscle and joint function in general. The information provided herein only serves as a guide for future investigations.

REFERENCES

1. Bigland, B. and Lippold, O.C.J., "The relation between force, velocity and integrated electrical activity in human muscle", *J. Physiol.*, **123**: 214–224 (1954).
2. Close, J.R. and Kidd, C.C., "The functions of the muscles of the thumb, index and the long fingers", *J. Bone Joint Surg.*, **51A**: 1601–1620 (1969).
3. Close, J.R., Nickel, E.D. and Todd, F.N., "Motor unit action potential counts — Their significance in isometric and isotonic contractions", *J. Bone Joint Surg.*, **42A**: 1207–1222 (1960).

4. Forrest, W.J. and Basmajian, J.V., "Functions of thenar and hypothenar muscles", *J. Bone Joint Surg.,* **47A**: 1585-1594 (1965).

5. Hagberg, M. and Jonsson, B., "The relationship between myoelectric activity and muscle work in intermittent statical contractions", *Electromyography Clin. Neurophysiol.,* **15**: 253-258 (1975).

6. Hall, E.A. and Long, C. II, "Intrinsic hand muscles in power grip. An electromyographic study", *Electromyography,* **8**: 377-421 (1968).

7. Inman, V.T., Ralston, H.J., Saunders, J.B., Feucstein, B and Wright, E.W., "Relations of human electromyogram to muscular tension", *Electroencepholography Clin. Neurophysiol.,* **4**: 187-194 (1952).

8. Landsmeer, J.M.F., "Power grip and precision handling", *Ann. Rheum. Dis.,* **21**: 164-170 (1962).

9. Lippold, O.C.J., "The relation between integrated action potentials in human muscle and its isometric tension", *J. Physiol.,* **117**: 492-499 (1952).

10. Long, C. II and Brown, M.E., "Electromyographic kinesiology of the hand. Muscles moving the long finger", *J. Bone Joint Surg.,* **46A**: 1683-1706 (1964).

11. Nightingale, A., "Relationship between muscle force and EMG in stand-at-ease position", *Ann. Phys. Med.,* **V**: 187-191 (1960).

12. Perry, J., "Electromyography as a force measurement", *NIH Grant. Final Project Report,* (1974).

13. Ralston, H.J., Inman, V.T., Strait, L.A. and Shattrath, M.D., "Mechanics of human isolated voluntary muscle" *Am.J.Physiol.,* **151**: 612-620 (1947).

14. Boivin, G., Wadsworth, G.E., Landsmeer, J.M.F. and Long, C.H., "Electromyographic kinesiology of the hand: Muscles driving the index finger", *Arch. Phys. Med.,* **50**: 17-26 (1969).

15. Napier, J.R., "The prehensile movements of the human hand", *J. Bone Joint Surg.,* **38B**: 902-913 (1956).

16. Bouisset, S. and Gouble, F., "Integrated electromyographical activity and muscle work", *J. Appl. Physiol.,* **35**: 5: 695 (1973).

17. Bouisset, S. and Maton, B., "Quantitative relationship between surface electromyography and intramuscular electromyographic activity in voluntary movement", *Am. J. Phys. Med.,* **51**: 285-295 (1972).

18. Gans, B.M. and Noordergraag, A., "Voluntary skeletal muscles: A unifying theory on the relationship of their electrical and mechanical activities", *Arch. Phys. Med. Rehabil.,* **56**: 194 (1975).

19. Jonsson, B. and Reichman, S., "Reproducibility in kinesiologic EMG: Investigations with intramuscular electrodes", *ACTA Morph. Neerl. Scand.,* **7**: 73-90 (1968).

20. Perry, J., Antonelli, D. and House, K., "The influence of velocity on electromyography as a force measurement", *Trans. 22nd Annual Mtg. Orthopaedic Research Society*, **1**: 238 (1976).

21. Zuniga, E. and Simon, D.G., "Nonlinear relationship between averaged electromyogram potential and muscle tension in normal subjects", *Arch. Phys. Med.*, **50**: 613 (1969).

22. Komi, P.V. and Buskirk, E.R., "Effect of eccentric and concentric muscle conditioning on tension and electrical activity of human muscle", *Ergonomics*, **15**: 417–434 (1972).

23. Metral, S. and Cassan, G., "Relationship between force and integrated EMG activity during voluntary isometric anisotonic contraction", *Europ. J. Appl. Physiol.*, **46**: 185–198 (1981).

24. Basmajian, J.V., *Muscles Alive. Their Functions Revealed by Electromyography,* (Williams and Wilkins, Baltimore, 1967).

25. Basmajian, J.V., Clifford, H.C., McLeod, W.D. and Hunally, H.N., *Computerism Electromyography,* (Butterworths, London & Boston, 1975).

26. Basmajian, J.V. and Cross, G.L. III, "Duration of motor unit potentials from fine wire electrodes", *Am. J. Phys. Med.*, **3**: 144–148 (1971).

27. deDuca, C.J. and Forrest, W.J., "Force analysis of individual muscles acting simultaneously on the shoulder joint during isometric abduction", *J. Biomechanics,* **6**: 385–393 (1973).

28. Gydikov, A., Dimitrov, D., Kosarov and Dimitrove, N., "Functional differentiation of motor units in human opponens pollicis muscle", *Exp. Neurology,* **50** (1976).

8

CAPSULOLIGAMENTOUS STRUCTURES OF THE METACARPOPHALANGEAL JOINT AND THEIR ROLE IN JOINT STABILITY

INTRODUCTION

The metacarpophalangeal (MCP) joint is a diarthroidial joint that permits flexion-extension and abduction-adduction as well as axial rotation of the proximal phalanx on the metacarpal head. Joint stability with which to restrain isometric functional forces is provided by the joint articular surfaces, capsuloligamentous structures and musculotendinous units. In general, the muscle and tendon forces are assumed to be the primary joint stabilizers. The ligaments and capsules provide initial stability, as well as a second line of defense[1].

Descriptive and qualitative anatomy of the capsuloligamentous structures and volar plate have been reported in the literature[2-9]. Landsmeer[9] described in detail the appearance of the collateral ligaments. He noted that at the MCP joint, the ligaments have their greatest obliquity to the longitudinal axis of the metacarpal at the index finger, with progressively less obliquity in the long, ring and little fingers. The ulnar collateral ligaments (UCL) are less oblique than the radial collateral ligaments (RCL) and smaller in size as well. For each ligament, due to the eccentric location of its origin relative to the axis of joint rotation, its length and thus the tension are changing as a function of the joint flexion angle.

Quantitative description of the capsuloligamentous structures in resisting joint force has been limited[8,10,11,12]. Zancolli, in a detailed treatise on hand functional anatomy, used cross-sectional analysis of the MCP joint to describe the concept of an "assemblage" or "force nucleus" which represented the site of action of forces across the MCP joint[10]. Subluxation and ulnar drift were explained by this model. The roles of the metacarpoglenoidal ligament as a support for the flexor tendon and metacarpophalangeal ligaments in resisting subluxation of the proximal phalanx during flexion were examined by Flatt and Fischer[11]. From a distraction test, the UCL was found to be stiffer than the RCL. James et al.[8] showed that an increase of intra-articular pressure facilitated ulnar deviation deformation.

In this chapter, the detailed morphometric data of the collateral ligaments around the MCP joint are presented. This information was collected by gross dissection and extra- and intra-articular observation of the ligaments. The locations of the origins and insertions of the ligaments and the variation of the ligament lengths throughout the range of joint motion were obtained by using biplanar radiographic technique. Furthermore, the relative contributions of each capsuloligamentous element to the overall joint constraint and stability were examined by using load-displacement tests on a universal testing machine. The application of the results in prosthetic design and joint reconstructive procedures are discussed.

MATERIALS AND METHODS

A. Gross Examination

The morphology of the capsuloligamentous structures was studied by gross dissection of cadaver finger specimens and observed with 2.5X loupe magnification. The locations of the origins on the metacarpal head and the insertions on the base of the proximal phalanx were identified. In addition, the structure was also examined from the internal surface. This was accomplished by splitting the finger longitudinally and removing the metacarpal head. From the interior, the trajectories of the fibers in various angulations were studied.

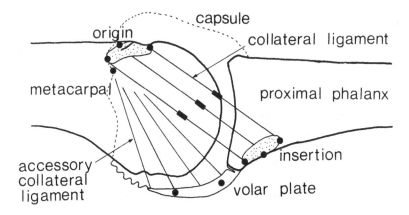

Fig. 8–1 Placement of metal markers in the ligamentous structures. Metal markers were inserted precisely at the origin, midportion and insertion sites for the dorsal, middle and volar thirds of the collateral ligaments and at the origin and insertion sites of the accessory collateral ligaments.

B. Biplanar Radiography

In order to quantify the locations of the origins and insertions of the capsuloligamentous structures and to obtain the changes of ligament lengths as a function of the joint flexion/extension angle, the biplanar X-ray technique was used. Metal markers were inserted at origin and insertion sites of the dorsal, middle and volar thirds of both collateral ligaments. Additional markers were inserted at the midportions near the joint line. Markers were also inserted at both the origin and insertion of the accessory collateral ligament on the cartilaginous portions of the volar plate (Fig. 8-1). Biplanar X-rays were then taken in both the AP and lateral planes with the joint flexed at various flexion angles with the articular surfaces kept in contact by a five-pound load.

The locations of the origins and insertions were then obtained by digitizing the X-ray films based on an established coordinate system (Fig. 8-2). The lengths of ligaments at various joint flexion angles were calculated based on the distance between these representative markers.

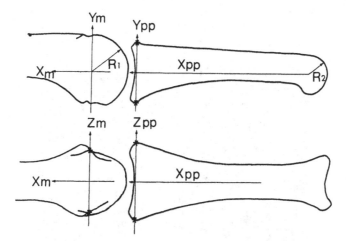

Fig. 8-2 Definition of coordinate system. Lateral view (top) and anteroposterior view (bottom) of the index MCP joint. The subscripts: m = metacarpal, pp = proximal phalanx, R_1 and R_2 are the radii of the metacarpal head and the head of the proximal phalanx, respectively. Origins of the metacarpal and proximal phalangeal coordinate systems bisect the tubercles of the metacarpal head and proximal phalanx, respectively.

C. Load-Displacement Test

In order to quantitate the amount of constraint provided by the ligamentous structures in resisting displacement of the joint and to examine the relative contribution of these elements to joint stability, load-displacement testing of cadaver finger specimens was performed.

Finger specimens were disarticulated proximally at the CMC joint and distally at the PIP joint. Skin, subcutaneous tissues and tendons were removed preserving the capsuloligamentous system. Two sets of 0.625mm K-wires oriented at right angles were inserted at both the proximal end of the metacarpal and the distal end of the proximal phalanx. The ends of the specimens with the K-wires were then embedded in bone cement and fastened to a universal testing machine with careful alignment in a true anteroposterior plane (Fig. 8-3).

Four modes of rotation and translational displacement, with reference to the proximal phalanx, were performed (Fig. 8-4). For each specimen, the tests were conducted in two parts. First, with the capsuloligamentous structures intact, the specimens were tested under all four modes of displacement at three joint flexion angles ($0°$, $45°$ and $75°$). Following this test, the specimens were divided into two groups for testing with sequential sectioning; one for both distraction and pronation-supination and the other group for volar-dorsal displacement and abduction-adduction. Each of the two groups was further divided into two subgroups for testing at two joint positions, $0°$ and $75°$. The sectioning of soft tissue was performed with a surgical blade. After sectioning

Fig. 8-3 Finger specimen mounted on the MTS universal testing machine for load-displacement test.

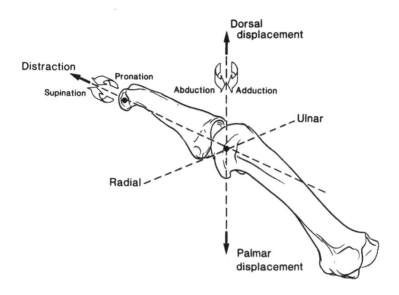

Fig. 8-4 Four modes of rotational and translational displacement were tested with the proximal phalanx as a reference.

of one element, the test was repeated under the same prescribed displacement. From the recorded load-displacement curves, the laxity and stiffness of the joint were obtained. The relative contribution of each ligament was also calculated by measuring the difference in the force or moment pre- and post-sectioning at a given displacement.

RESULTS

A.   *Morphology*

The collateral ligaments were separable into two layers; one superficial and the other deep (Figs. 8-5a and 8-5b). The origin of the RCL is more distal than that of the UCL and the obliquity of the UCL is less than that of the RCL. Both collateral ligaments arise from a crescent or comma-shaped area on the tubercle of the metacarpal. These concave arcs lie dorsal and proximal to the axis of rotation with the result that the collateral ligaments function eccentrically during joint movement.

The sites of origin and insertion of both collateral ligaments were obtained by averaging the results of seven specimens (Table 8-1 and Figs. 8-6a and 8-6b). On the AP view, the origins of the RCL are closer to the center line of

Table 8-1 Locations of origins and insertions of the ligaments in a representative finger model. (n = 7)

Ligaments		Locations (mm) of					
		Origin (metacarpal)§			Insertion (proximal phalanx)ζ		
		X	Y	Z	X	Y	Z
RCL†	Dorsal portion	−0.8 + 0.3*	7.2 + 0.3	−1.4 ± 0.5	−3.2 ± 0.2	−3.5 ± 0.3	−6.5 ± 0.3
	Middle portion	3.1 + 0.5	7.9 + 0.4	−3.7 ± 0.6	−2.3 ± 0.7	−5.0 ± 0.3	−5.2 ± 0.7
	Volar portion	4.6 + 0.6	6.6 + 0.4	−5.3 ± 0.3	−2.5 ± 0.4	−5.3 ± 0.4	−4.4 ± 0.3
UCL	Dorsal portion	0.0 ± 0.4	6.0 ± 0.4	6.1 ± 0.6	−3.4 ± 0.2	−3.1 ± 0.6	7.2 ± 0.7
	Middle portion	2.5 ± 0.6	6.3 ± 0.5	7.1 ± 0.5	−2.6 ± 0.2	−4.4 ± 0.4	7.1 ± 0.5
	Volar portion	4.8 ± 0.7	6.6 ± 0.5	6.8 ± 0.6	−0.6 ± 0.8	−4.6 ± 0.9	6.4 ± 0.8
RACL	Middle portion	4.2 ± 0.2	5.3 ± 0.2	−5.6 ± 0.2	- -	- -	- -
UACL	Middle portion	4.1 ± 0.4	4.8 ± 0.2	6.6 ± 0.3	- -	- -	- -

* Mean ± standard deviation

§ Measured from the metacarpal coordinate system

ζ Measured from the proximal phalanx coordinate system

†RCL = radial collateral ligament,   UCL = ulnar collateral ligament,   RACL = radial accessory collateral ligament, and
UACL = ulnar accessory collateral ligament

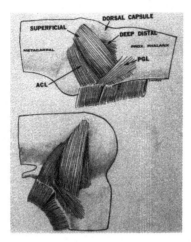

Fig. 8-5 Morphological findings of the ligamentous structures of the MCP joint.

(a) External view. Extended position (top) and flexed position (bottom). The collateral ligament appears to be separable into two components; the superficial and the deep. The deep distal component is hidden beneath the superficial portion of the collateral ligament when the joint is in extension but becomes increasingly taut and visible as the joint is flexed. ACL = accessory collateral ligament; PGL = phalangioglenoidal ligament.

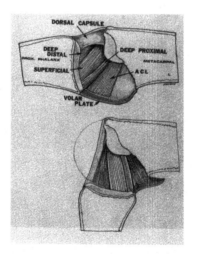

Fig. 8-5 Morphological findings of the ligamentous structures of the MCP joint.

(b) Internal view. Extended position (top) and flexed position (bottom). The deep layer of the collateral ligament appears to be separable into two portions; the deep distal and the deep proximal. The deep proximal component shortens in flexion and lengthens in extension. These fibers bulge laterally during joint flexion. ACL = accessory collateral ligament.

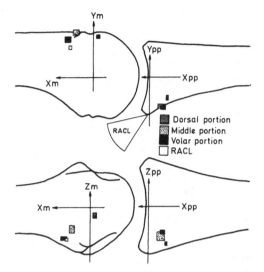

Fig. 8–6(a)   Locations of origins and insertions of the radial collateral ligament and radial accessory collateral ligament. Lateral view (top) and anteroposterior view (bottom).

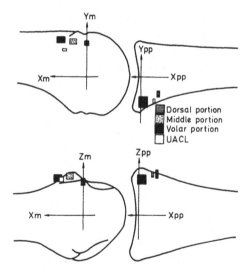

Fig. 8–6(b)   Locations of origins and insertions of the ulnar collateral ligament and ulnar accessory collateral ligament. Lateral view (top) and anteroposterior view (bottom).

the metacarpal than that of the UCL. The accessory collateral ligament has a moving insertion associated with the angulation of the proximal phalanx during MCP flexion-extension. This is because the ligament inserts on the cartilaginous portion of the volar plate which bends from 10° to 55° with respect to the metacarpal shaft during MCP joint flexion.

B. Lengths of Collateral Ligaments

Due to the eccentric locations of the origins about the axis of joint rotation, the fibers of the collateral ligaments change their apparent lengths during joint rotation. The variations of the fiber lengths, as represented by the distance between the origin and insertion markers, were obtained by using biplanar radiography throughout the range of MCP joint flexion (Figs. 8–7a and 8–7b). When the joint flexed from 0° to 80°, the dorsal portions elongated 3 to 4mm and the middle 0.5 to 1mm. The volar portion, however, shortened 1 to 2mm. The variation of the ligament lengths during abduction-adduction rotation was not studied experimentally but was calculated based on the locations of the origins and insertions (Fig. 8–8).

C. Restraining Forces and Moment

When the joint is displaced either in rotation or translation, the load displacement curves are usually nonlinear. According to the literature, these curves can be approximated by three linear segment[13] (Fig. 8–9). The slope of the line at the middle inflection point was defined as "neutral stiffness". At the end of the curve, the corresponding slopes are called "terminal stiffness". The laxity is defined as the distance between the intersection of these three lines. Joint laxity and terminal stiffness of the intact specimen under four modes of displacement are summarized in Tables 8–2 and 8–3. The laxities of

Table 8–2 Joint laxity in each mode of displacement (n = 15)

Joint Flexion Angle	Translational (mm)			Mode of Displacement	Rotational (deg)		
	Dist.	Dorsal	Volar	Pro.	Sup.	Abd.	Add.
0	4 ± 1	10 ± 1	3 ± 1	26 ± 4	25 ± 6	10 ± 2	11 ± 2
45	2 ± 0	8 ± 1	3 ± 0	21 ± 3	23 ± 6	8 ± 3	7 ± 1
75	2 ± 0	5 ± 1	3 ± 1	18 ± 3	18 ± 3	3 ± 1	4 ± 1

* Mean ± Standard Error

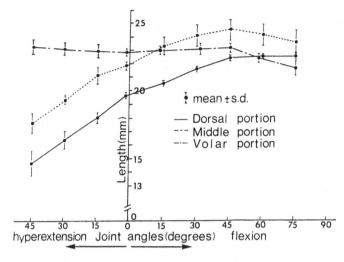

Fig. 8-7 Variation of the distance between origin and insertion of the collateral ligaments as a function of joint flexion angle.

(a)   Radial collateral ligament. Dorsal and middle portions shorten during extension (hyperextension) and lengthen in flexion. The volar portion length remains nearly constant, shortening at maximum flexion.

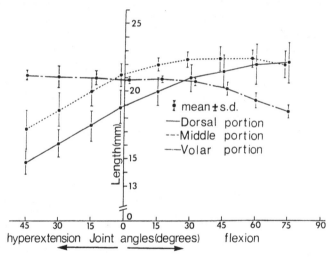

Fig. 8-7 Variation of the distance between origin and insertion of the collateral ligaments as function of joint flexion angle.

(b)   Ulnar collateral ligament. Dorsal and middle portion retain similar length changes with flexion, while volar ligament shortens after 30° of flexion.

Table 8-3 Terminal stiffness in each mode of displacement (n = 15)

Joint Flexion Angle	Mode of Displacement: Translational (10^4 N/m) Dist.	Dorsal	Volar	Rotational (10^{-2} N m/deg) Pro.	Sup.	Abd.	Add.
0	15 ± 4	3 ± 0	6 ± 1	5 ± 1	4 ± 1	12 ± 1	12 ± 1
45	11 ± 3	4 ± 0	4 ± 1	5 ± 1	5 ± 1	18 ± 5	21 ± 4
75	7 ± 1	3 ± 1	5 ± 0	6 ± 1	6 ± 1	21 ± 3	22 ± 2

* Mean ± Standard Error

all modes of displacement of the proximal phalanx decreased with increasing flexion angles.

The relative contributions of the capsuloligamentous structures in resisting the four modes of joint displacement were obtained from the sequential sectioning test (Figs. 8-10 to 8-13). In these graphs, the amount of decrease in

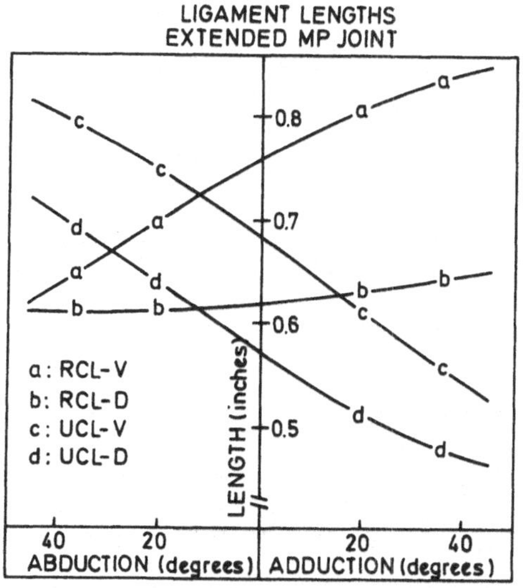

Fig. 8-8 Changing lengths in the radial and ulnar collateral ligaments during abduction and adduction of the MCP joint in the extended position. Note the greater length change (laxity) in the radial collateral ligament (dorsal portion) in comparison to the ulnar ligament system.

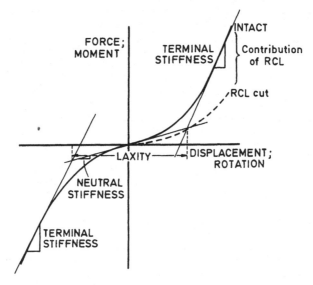

Fig. 8-9 Typical joint load-displacement curves are usually nonlinear and can be approximated by three segments of straight lines.

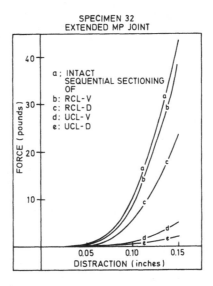

Fig. 8-10 Distraction test. The amount of decrease in constraint forces between curves represented the contribution of that particular element being sectioned. Both the RCL and UCL provided primary constraints in resisting distraction displacement.

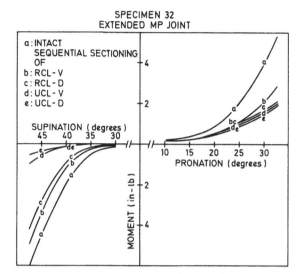

Fig. 8-11 Pronation-supination test. The amount of decrease in constraint moments between curves represented the contribution of that particular element being sectioned. RCL and UCL are primary constraints for pronation and supination, respectively.

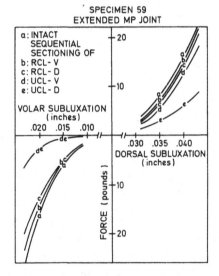

Fig. 8-12 Volar-dorsal displacement test. The amount of decrease in constraint forces between curves represented the contribution of that particular element being sectioned. The UCL was found to provide primary constraints in both volar and dorsal displacement.

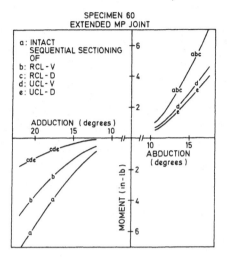

Fig. 8-13 Abduction-adduction test. The amount of decrease in constraint moments between curves represented the contribution of that particular element being sectioned. The UCL and RCL provided major constraints in resisting abduction and adduction motion, respectively.

constraint force or moment between curves represented the contribution of that particular element marked as being cut.

As anticipated, the combination of RCL and UCL contributed about 90 percent and 70 percent of the distraction force at 0° and 75° flexion angles, respectively. The volar portions of both ligaments contributed more at the extended position (Fig. 8-10). At the flexed position, the dorsal portions became more important. The UCL was stiffer than the RCL at both 0° and 75° flexed positions.

To resist pronation of the proximal phalanx, the RCL contributed 50 percent to 70 percent (Fig. 8-11). In the extended position, the relative contribution was equally provided by dorsal, volar and accessory portions. However, at 75° flexion the dorsal portion provided the major resistance. The dorsal capsule contributed only 10 percent at both flexed positions.

For supination, the primary constraints were provided by the UCL. The dorsal capsule contributed more than ten percent throughout the range of flexion, while the volar plate contributed minimally.

The forces to resist volar displacement of the proximal phalanx were almost equally shared by the RCL and UCL in the extended position (Fig. 8-12). However, in the flexed position the UCL contributed almost twice as much as the RCL. The contribution from the accessory collateral ligament, volar plates and dorsal capsules was negligible.

For resisting dorsal displacement, the primary constraints were the dorsal portions of the RCL and UCL. Both the volar plates and the dorsal capsules provided moderate constraints.

In abduction and adduction, the restraint moments were provided almost entirely from the RCL and UCL, respectively, throughout the range of flexion (Fig. 8–13). The accessory collateral ligaments were important in resisting this mode of displacement when the joint was in an extended position.

DISCUSSION

The collateral ligaments appear to be somewhat more complex in structure than noted in previous descriptions. If observing the ligament from the outer surface, the directions of the fibers appear to be essentially parallel as they pass distovolarly to the base of the proximal phalanx. However, if one inspects the system from its deep surface closer to the joint, distinct cleavages are apparent, and the partially spiral nature of the deep ligament can be appreciated.

As noted by Hagert[6], those fibers farthest from the center of rotation undergo the greatest change in length, and this eccentricity is crucial to stability at progressive angulations of the joint. The results of the change of ligament lengths during motion suggest that the dorsal portions of both collateral ligaments will provide the main restraints when the joint is flexed, while the volar portions are more important when the MCP joint is extended. The data of ligament lengths at various joint configurations are essential to understanding the mechanism of extension contractures at the MCP joint from burns and dystrophy. The results also confirm the need to immobilize the MCP joint at 50° and 70° flexion to prevent extension contractures.

Precise evaluation of the relative contribution of the capsuloligamentous structures to the stability of the MCP joint has not previously been performed. Joint laxity in all four modes of displacement uniformly decreased with an increase of the flexion angle, except in distal distraction. Among the three modes of translation displacement, the laxity in dorsal displacement was the greatest. This is due to the obliquity of the collateral ligaments and the minimum stabilizing role of the volar plates. Surprisingly, a large amount of laxity was observed in pronation-supination displacement as compared to those ranges of motion *in vivo*. This might suggest that the skin and subcutaneous tissue play an important role in stabilizing axial rotation. In addition, the extensor tendon complex with its attachments to the transverse intermetacarpal ligament and flexor retinaculum through the sagittal bands probably has a significant role as well.

In resisting distraction, the UCL has been noted to contribute more than the RCL[8]. This was also confirmed by the present study. Although the RCL is longer and thicker than the UCL, the UCL became tauter earlier in distraction due to its less obliquity.

The accessory collateral ligaments (ACL) play a significant role in restraint of abduction-adduction in spite of their attachment to the lateral border of the mobile cartilaginous portion of the volar plate. Restoration of ACL integrity during joint reconstruction should be considered.

REFERENCES

1. Basmajian, J.V., "The unsung virtues of ligaments", *Surg. Clin. No. Am.*, **54**: 1259–1267 (1974).
2. Eaton, R.G., *Joint Injuries of the Hand,* (Springfield, Chas. C. Thomas Co., 1971).
3. Flatt, A.E., "Fracture-dislocation of an index metacarpophalangeal joint", *J. Bone Joint Surg.*, **48-A**: 100–104 (1966).
4. Flatt, A.E., "The pathomechanics of ulnar drift. A biomechanical and clinical study", *Final Report, SRS Grant RD 2226M* (1971) pp. 1–145.
5. Glad, P., "The anatomy of the volar parts of the capsules of the finger joints", *J. Bone Joint Surg., 49-A:* 362–367 (1967).
6. Hagert, C.G., "Anatomical aspects on the design of metacarpophalangeal implants", *Reconstr. Surg. Traumat.*, **18**: 92–110 (1981).
7. Hakstian, R.W. and Tubiana, R., "Ulnar deviation of the fingers: The roles of joint structure and function", *J. Bone Joint Surg.*, **49-A**: 299–316 (1967).
8. James, D.F., Clark, I.P., Colwill, J.C. and Halsall, A.P., "Forces in metacarpophalangeal joint due to elevated fluid pressure — Analysis, measurements and relevance to ulnar drift", *J. Biomech.*, **15**: 73–84 (1982).
9. Landsmeer, J.M.F., "The coordination of finger-joint motions", *J. Bone Joint Surg.*, **45-A**: 1654–1662 (1963).
10. Zancolli, E., *Structural and dynamic bases of hand surgery*. (Philadelphia, J.B. Lippincott, 1968).
11. Flatt, A.E. and Fischer, G.W., "Restraints of the metacarpophalangeal joints: A force analysis", *Surg. Forum.* **19**: 459–460 (1968).
12. Berme, N., Paul, J.P. and Purves, W.K., "A biomechanical analysis of the metacarpophalangeal joint", *J. Biomech.*, **10**: 409–412 (1977).
13. Markolf, K.L., Mensch, J.S. and Amstutz, H.C., "Stiffness and laxity of the knee. The contributions of the supporting structures", *J. Bone Joint Surg.*, **58-A**: 583–594 (1976).

9

EPILOGUE

The mechanics of the human hand is an important basic medical science which is required to understand hand function and to treat debilitating hand problems. The hand is an intricate extension of the mind's activity and creativity, a specialized tool mounted on a multijointed lever arm, whose purpose is to reflect in writing, designing, building, manipulating, machining, and other prehensile actions, the concepts and demands of individual thought. What engineer, surgeon, musician, or artist, let alone machinist, or farmer, does not recognize the utmost importance of a functional hand unit, to be productive in today's society. Each year, unfortunately, hand injuries occur at an alarming rate, estimated at 16 million upper extremity hand-wrist injuries, with two-thirds in individuals still in their working years[1]. In addition, debilitating disease processes, such as rheumatoid arthritis and degenerative arthritis, further incapacitate the hand, limiting not only work activity but the ability to perform basic tasks of daily living[2].

Reconstruction of injured tendons and joint ligaments, fracture fixation, and many other operative, as well as nonoperative, treatments of hand problems are practiced by over ten thousand hand surgeons daily. Finger joint replacement with silastic, cemented, and porous ingrowth prostheses represents an estimated ten-million-dollar per year medical health investment. Nonetheless, fundamental understanding of the basic biomechanical principles on which hand reconstructive surgery and joint replacement are based is lacking, and the bench research needed to answer questions concerning hand function and its mechanical equivalent has been met by only a few investigative teams.

For the past twenty years, collaborative efforts between engineers and hand surgeons have developed and grown in an attempt to explain the basic mechanical equivalents of finger and thumb joint functions. Research plans have been devised to develop analytical models to understand normal hand function and apply that knowledge to injury or diseases that affect hand function. We have expanded on the works of Bunnell[3], Landsmeer[4], Kaplan[5], Flatt[6], and others who recognized the importance of hand biomechanics in order to understand the complexities of a multijointed, cantilevered system of tendons, muscles, and ligaments which act to provide strength, motion, and dexterity to the human hand.

In applying our biomechanical data to develop a finger joint prosthesis, more questions have been raised than answered. The complexities of finger and thumb joints have not allowed simple biologic or mechanical solutions to implant design, fabrication and clinical acceptance. Nonetheless, analysis of a number of the different elements which interact to produce a hand which is not only mobile but stable during pinch and grasp has been accomplished. This model has been used to generate the results contained in this book.

Two unstudied areas are the extensor tendon mechanism and the flexor pulley system. Their mechanical effect on the hand and interplay with both joint force transmission and flexor tendon function is significant. Both systems affect normal balance and equilibrium of analytic hand models as well as the direction and magnitude of joint forces and moments. Clinically, abnormalities in the extensor tendon mechanism and flexor pulley system have been treated by a "seat-of-the-pants" approach to reconstruction, with ignorance of fundamental anatomic as well as mechanical principles of function. To complete analytic studies and to develop an accurate normative model, the extensor tendon mechanism and flexor tendon pulley system must be introduced and studied as part of the hand force system.

The contribution of the extensor tendons to hand motion, force and balance is the least understood area of hand biomechanics. When extensor tendon force is lost, finger function is greatly affected. Anatomically, the extensor tendons and intrinsic muscles interdigitate across three finger and two thumb joints and function as a single fascial tendon expansion, with firm attachments to fixed points on each joint. The tendon lengths between these points of fixation have interposed muscle units and function by means of precise tendon length units. This is very different from the flexor system, which has an independent flexor for each joint. Controversy continues as to whether the transverse lamina controls the dorsal-volar movement of the lateral bands[7, 8, 9] or whether the distal tendon insertion, which moves up and down, changes the dorsal and volar migration of the lateral bands[10, 11]. Force distribution between these tendon slips has been studied to a certain extent through trigonometric diagrams[12] and direct force measurement with strain gauge device[13]. In our previous studies, the relationship of force distribution among each of the components of the extensor mechanism was found to be of paramount significance to the overall muscle force determination of the hand. Experimental verification and analytic modeling of these relationships is urgently required. Such information is also important for clarifying the deformities of finger and thumb joints in disease processes such as rheumatoid arthritis. Effective extensor tendon repair techniques and better hand function should result from such studies.

The flexor tendon pulley system is an equally significant and yet not well understood component of hand force transmission. The flexor pulley system acts to maximize tendon excursion for mobility yet control tendon moment arms during pinch and grasp. Despite work by Doyle[14], Hunter[15] and Manske[16], the importance of the flexor pulley system to finger pinch and grasp strength and coordinated hand motion has been overlooked. Tendon excursion to produce full finger and thumb motion has been accepted by most hand surgeons as the goal of reconstructive tendon surgery. The mechanical consequences of pulley release and incomplete pulley repair remain unknown and the proper placement of pulleys, the material and methods of reconstruction and the effect on hand strength associated with failure to repair pulleys have not been studied. We propose to resolve the pulley force distribution by analytic and mechanical studies which will provide a sound basis to study pulley repair and reconstruction.

A further important physiological constraint which has not been considered in the model for tendon force analysis is the relationship between tension and length of the muscles. This well-recognized physiological relationship is not only significant in the proper determination of muscle forces under normal conditions but is also of paramount importance in pathological conditions where the normal relationship between the joint and tendon excursion has been altered. Information in terms of muscle morphology and its relationship to muscle tension has been summarized in this book and reported elsewhere[17, 18]. With the inclusion of such a relationship in the analytic model, we believe that hand force analysis and resultant joint and tendon forces can be more realistically predicted.

A final area of interest and significance is thumb biomechanics. Analytic force studies have been performed with estimations of joint and tendon forces. However, these studies lack ligament-capsule force assessment. Experimental and clinical verification of our normative and pathologic models of the thumb are needed. We have developed a special s-shaped force transducer to measure tendon tension during carpal tunnel surgery for *in vivo* validation of analytically predicted tendon forces[19]. The measured values are within the range of forces predicted for the thumb, FPL, index, FDS and FDP, and long finger, FDS, in pinch function with different strengths. These encouraging results must be expanded to verify our theoretical models for finger and thumb, tendon and joint force determination. The thumb metacarpophalangeal (MCP) joint will be of specific interest since it has a high incidence of injury (skier's thumb) and the thumb basal joint which develops degenerative arthritis, secondary to chronic ligament instability and excessive joint loading.

By extending well-developed techniques of ligament force and stiffness measurement to the thumb, one can apply such new information to the pathophysiology of these injuries, associated disease processes and their treatment.

Kinematics of the joint are controlled by the contour of articular surfaces, the constraints of capsuloligamentous structures and the contraction of muscles across the joint. In the preliminary results summarized in this book, the intricate relationship between the articulating kinematics and the constraints of surrounding ligaments have been well demonstrated. Further studies of these relationships will provide renewed understanding of the mechanisms of joint stability and the etiology of arthritic deformities. Clinical questions regarding MCP joint contracture and the design criteria for resurfacing prostheses could then be analyzed.

The ultimate significance of this scientific endeavor is to direct hand biomechanics postulates to clinically soluble hand problems. We started our investigations in this direction with gratifying results to provide insight in neuropathic hand deformity, tendon laceration pathology, finger joint ligament contracture, and rheumatoid hand deformity. We anticipate that investigations into hand forces, both normal and pathologic, will continue to be fruitful by establishing basic principles of hand function and recommendations for improved treatment of hand injuries and disease. Our colleagues in hand surgery, hand therapy and biomechanics have been responsive to and motivated by the materials and methodology presented in this book. We sincerely hope that the light of hand biomechanics as a clinical tool is beginning to glimmer at the end of the tunnel. What we have accomplished represents the mere beginning of a most challenging and rewarding endeavor which will require dedicated lifetime efforts by many hand surgeons, medical scientists, and biomechanicians.

BIBLIOGRAPHY

1. American Academy of Orthopaedic Surgeons *Musculoskeletal System Research-Current and Future Research Needs.* (AAOS Publications, Chicago, 1981).

2. National Commission on Arthritis and Related Musculoskeletal Disease, *Report to the Congress of the United States.* vol. I. The Arthritis Plan, pp. 100-102, April 1976.

3. Bunnell, S., *Surgery of the Hand.* (J. B. Lippincott Company, Philadelphia, PA, 1944).

4. Landsmeer, J.M.F., "Anatomical and functional investigations on the articulation of human fingers", *Acta Anat. Suppl.* **24** (1955).

5. Kaplan, E., *Functional and Surgical Anatomy of the Hand.* (J.B. Lippincott Co., Philadelphia, 1965).

6. Flatt, A.E., "The pathomechanics of ulnar drift", *Social and Rehabilitation Services, Final Report.* Grant. No. RD 2226M (1971).

7. Landsmeer, J.M.F., "The anatomy of the dorsal aponeurosis of the human finger and functional significance", *Anat. Rec.,* **104**: 31-44 (1949).

8. Kaplan, E., "Anatomy, injuries and treatment of the extensor apparatus of the hand and the digits", *Clin. Orth.,* chapter 3, pp. 24-41 (1959).

9. Tubiana, R. and Valentin, P., "Physiology of the extensors of the fingers", *Surg. Clinics of N. Am.,* **44**: 908-910 (1964).

10. Harris, C. and Rutledge, G.L., "Functional anatomy of the extensor mechanism of the finger", *JBJS,* **54A**: 720 (1972).

11. Zweitin, K.J. Van, "The extensor assembly of the finger in man and non-human primates".(Univ. of Leiden, Doctoral Thesis, 1980).

12. Baumann, J.A., and Patry, G., "Observations microscopiques surla texture fibreuse de l'emsemble tendineux extenseur du doigt de la mam. Chez l'homme", *Revue Medicale de la Susisse Romande* **63**: 900-912 (1943).

13. Sarrafian, S.K., Kazarian, L.E., Topouzian, L.K., Sarrafian, Y.K. and Siegelman, A., "Strain variation in the components of the extensor apparatus of the finger during flexion and extension", *JBJS,* **52A**: 980-990 (1970).

14. Doyle, J.R. and Blythe, W., "The finger flexor tendon sheath and pulleys: Anatomy and reconstruction", *American Academy Ortho. Surg., Symposium on Tendon Surgery.* (St. Louis, The C.V. Mosby Company, 1975).

15. Hunter, J.M., Schneider, L.H., Mackin, E.J. and Bell, J.A., *Rehabilitation of the Hand.* (The C.V. Mosby Company, St. Louis, 1978).

16. Manske, P.R. and Lesker, P.A., "Palmar aponeurosis pulley", *J. Hand Surg.,* **8**(3): 259-263 (1983).

17. Otten, E., "Morphometrics and length — force relations of muscles", *Proceedings of IX International of Biomechanics,* p. 52, (Waterloo, Ontario, Canada Congress, (1983).

18. Woittiez, R.D., Huijing, P.A. and Rozendal, R.H., "Influence of muscle architecture on the length-force diagram: A model and its verification", *Pflugens Archiv, European Journal of Physiology,* **397**: 73-74 (1983).

19. Cooney, W.P., An, K.N., and Chao, E.Y.S., "Direct measurement of tendon forces in the hand", p. 53, *Trans. of 32nd Ann. Meeting, Orthopedic Research Society,* (New Orleans, Louisiana, February 17-20, 1986).

AUTHOR INDEX

SUBJECT INDEX

(*f* = figure; *t* = table;)

abduction-adduction strength, 114, 114*t*
abduction-adduction test, 176*t*
abductor digiti quinti, 33, 36
abductor pollicis brevis, 31-2, 36, 44, 47, 138
accessory adductor pollicis oblique, *see*
    adductor pollicis, accessory
    oblique head
ADD, *see* adductor pollicis
adductor pollicis, 32, 44, 48
    accessory oblique head, 32, 36, 41, 48
    volume, 41
ADQ, *see* abductor digiti quinti
angles, *see* Eulerian angles
antagonistic muscles, 35
anthropometric measurements, 108
anthropometric variations, 16
    minimization, 25
APB, *see* abductor pollicis brevis
Aristotle, v
arthritis, 74, 92
    degenerative, 73
    rheumatoid, 48, 73, 90
    see also osteoarthritis
axial rotation angle, 12, 13*f*, 80
    see also Eulerian angles

ball-and-socket configuration, 91
4-bar linkages, 34
basic feasible solution, 61, 64, 68
    see also feasible solution
Bell, Sir Charles, v, viii, 49
bi-articular system, 5
bipennate fashion, 32
biplanar radiographic technique, 41, 46, 74,
    165
    see also biplanar X-ray technique
biplanar X-ray technique, 6-7, 75*f*, 93, 97
    139*f*, 163

    see also biplanar radiographic technique
Blix curve, 49
bowstring (ing), 6, 48, 49
bowstring model, 27
Brand, P. W., vi
Bunnell, S., vi

capsuloligamentous structures, 163 — 178
carpal tunnel procedures, 69
carpometacarpal joint, 7
Cartesian coordinate axes, 79, 93
Cartesian coordinate systems, 9, 10*f*, 15-16
    54, 56*f*, 80
    distal (moving), 11, 12, 12*f*
    origins: center of concave articular
        surface, 16; center of rotation, 16
    primary, 9
    proximal (fixed), 11, 12, 12*f*
    secondary, 9
center of concave articular surface, 9, 16
center of rotation, 9, 16
centroid of muscle, 31
chuck grip, 33
chuck pinch, 106
CMC joint, *see* thumb
    carpometacarpal joint
collateral ligaments, *see under* ligaments
common perpendicular line, 78, 79
    magnitude, 94-5
constraint equations, 59, 61, 64
constraint forces, 53, 54
    joint, 68
constraint moments, 53, 54
coordinate systems, *see* Cartesian
    coordinate systems
coordinate transformations, 11
coordination, fingers and thumb, 74
correlation coefficients, 159

185